Computational Methods in Physics, Chemistry and Biology

Computational Methods in Physics, Chemistry and Biology

An Introduction

Paul Harrison

The University of Leeds, UK

JOHN WILEY & SONS, LTD

Chichester / New York / Weinheim / Brisbane / Singapore / Toronto

Other Wiley Editorial Offices

John Wiley & Sons, Inc., 605 Third Avenue,
New York, NY 10158-0012, USA

Wiley-VCH Verlag GmbH, Pappelallee 3,
D-69469 Weinheim, Germany

Jacaranda Wiley Ltd, 33 Park Road, Milton,
Queensland 4064, Australia

John Wiley & Sons (Asia) Pte Ltd, 2 Clementi Loop #02-01,
Jin Xing Distripark, Singapore 0512

John Wiley & Sons (Canada) Ltd, 22 Worcester Road,
Rexdale, Ontario M9W 1L1, Canada

Library of Congress Cataloging-in-Publication Data

Harrison, Paul
    Computational methods in physics, chemistry, and mathematical biology: an
Introduction / Paul Harrison
        p. cm
    Includes bibliographical references and index.
    ISBN 0 471 49562 X (cased) -- ISBN 047149563 4 (limp)
    1. Numerical analysis. 2. Mathematical physics.. I. Title

QC20.7. A5 H39 2001
519.4 – dc21   2001045412

British Library Cataloguing in Publication Data

A catalogue record for this book is available from the British Library

ISBN 0 471 0 471 49562 X (cased)   ISBN 047149563 4 (paperback)

Produced from Word files supplied by the author
Printed and bound in Great Britain by Antony Rowe Ltd, Chippenham, Wilts
This book is printed on acid-free paper responsibly manufactured from sustainable
forestry, in which at least two trees are planted for each one used for paper production.

To Evertonians

Contents

Preface

Since the earliest days of philosophical thinking, there have always been two ways to do science. The first is to go and out and observe and make measurements, for example dropping cannon balls from the Leaning Tower of Pisa, or studying the motion of the 'wanderers'. The second is by contemplating the results of the first, for example, our increase in apparent weight when a lift goes up. All scientific progress until the middle of the twentieth century was performed in this way and the achievements have been truly remarkable.

Now in the middle of the twentieth century, machines started to be developed that could perform arithmetical operations, and lots of them, very quickly. With our insatiable appetite to investigate the world around us and perform science, people quickly began to look at ways the new computers could be exploited in our quest for greater knowledge and understanding. And so a third way of doing science was created—computation.

The purpose of this book is to introduce, and hopefully inspire, young minds into exploiting these most valuable assets and forging ahead with scientific discovery in the twentyfirst century.

PAUL HARRISON

The University of Leeds

April 2001

Acknowledgements

I'd like to thank all those people who have believed in me and have helped (and still are helping) me along my way—they know who they are. Claire, Hannah and Joe for giving me a life and my parents for their genes (see Chapter 7).

I would also like to express my gratitude to my employers, The University of Leeds and the School of Electronic and Electrical Engineering, for providing me with an academic position with sufficient freedom for me to indulge myself in the things I like doing.

P.H.

About the author

The author is currently working in the Institute of Microwaves and Photonics, which is a research institute within the School of Electronic and Electrical Engineering at the University of Leeds in the United Kingdom. He can always be found on the web, at the time of writing, at

```
http://www.ee.leeds.ac.uk/homes/ph/
```

and always answers e-mail:

```
p.harrison@physics.org
```

Paul is working on a wide variety of projects, most of which centre around exploiting quantum mechanics for the creation of novel opto-electronic devices, largely, but not exclusively, in semiconductor *Quantum Wells, Wires and Dots*. Up to date information can be found on his web page. He is always looking for exceptionally well qualified and motivated students to study for a PhD degree with him—if interested, please don't hesitate to contact him.

About the book

100% of this book was produced with 'open source' not-for-profit software under the Linux operating system. The text was prepared with LaTeX2ϵ using Wiley's own style (class) files. It was input by hand with the aid of the superb 'Vi IMproved' (vim). The schematic diagrams were prepared using 'xfig' and the x-y plots with 'xmgr'. Thanks to all those contributing to the open software movement——I couldn't have done it without you.

Introduction

As stated in the preface, although the main theme of this work is clearly computational methods, in an attempt at completeness, the book makes a low-key start with a revision of fundamental ideas of classical mechanics and hence their analogies in quantum mechanics. The aim of this approach is to provide a refresher and hence to gear up the student to the point where the underlying physics knowledge can be taken as read before the new computational techniques are introduced.

The book was written with a single semester undergraduate course in computational physics in mind, though it is hoped that it may appeal to the chemistry undergraduate who needs to know about molecular dynamics and the mathematician or biologist who is studying population dynamics. It may also serve as a refresher for graduate students moving into computational science.

For more information, to download the computer sourcecodes and for help with the problems, see the book's website:

```
http://www.imp.leeds.ac.uk/CM/
```

1

Numerical Solutions to Schrödinger's Equation

1.1 PARTICLES AND CLASSICAL MECHANICS

The word 'particle' is taken to mean an entity which has mass, a definite position and perhaps a 'speed' or even 'velocity', basically an everyday object. The laws of physics which describe the behaviour of such objects have been know for hundreds of years and are often introduced in layers of increasing complexity. For example:

$$\text{speed} = \frac{\text{distance travelled}}{\text{time taken}} \qquad (1.1)$$

Definition of Speed

which is perhaps then expressed as:

$$v = \frac{s}{t} \qquad (1.2)$$

or even more exactly, the *average* speed:

$$\overline{v} = \frac{\Delta s}{\Delta t} \qquad (1.3)$$

where the delta symbols Δ are understood to mean the 'change in' distance s and the change in time t. In a similar way the idea of changing speed is introduced and called 'acceleration':

$$\text{acceleration} = \frac{\text{change in speed}}{\text{time taken for change}} \qquad (1.4)$$

Definition of Acceleration

or more mathematically:

$$a = \frac{\Delta v}{\Delta t} \qquad (1.5)$$

And from these humble beginnings, the familiar equations of motion (for constant acceleration in a straight line) are derived. For example, if the initial speed of a particle is u and a time t later it has changed to v, then:

$$a = \frac{v - u}{t} \tag{1.6}$$

from which:

$$at = v - u \qquad \therefore v = u + at \tag{1.7}$$

The distance travelled during such an acceleration would be:

$$\text{distance travelled} = \text{average speed} \times \text{time} \tag{1.8}$$

which in mathematical language would be written as:

$$s = \left(\frac{u + v}{2}\right) t \tag{1.9}$$

Using the expression for the final velocity v in equation (1.7) then:

$$s = \left(\frac{u + u + at}{2}\right) t \tag{1.10}$$

i.e.

*Distance
Travelled*

$$s = ut + \frac{1}{2}at^2 \tag{1.11}$$

A textbook example of motion along a straight line with constant acceleration is that experienced by falling particles in the Earth's gravitational field—something reputedly discovered by Galileo Galili when he dropped two cannonballs of different mass from the top of the Leaning Tower of Pisa. He observed that they hit the ground at the same time—thus dispelling two millennia of dogma—*the different masses fall at the same speed* (they accelerate at the same rate, call it g).

Another quantity of importance became apparent, the 'momentum'. This represented the impetus of a moving particle, i.e. a heavy horse and carriage was much more difficult to stop than an air-filled pig's bladder travelling with the same speed. The former was therefore said to be travelling with more momentum. Furthermore, the faster the horses trotted or cantered, the more momentum the horse and carriage had. This led to the mathematical definition:

*Definition of
Momentum*

$$\text{momentum} = \text{mass} \times \text{speed} \tag{1.12}$$

(note, still only considering motion along a straight line). Or symbolically:

$$p = mv \tag{1.13}$$

The development of calculus allowed these ideas to be expressed with much more mathematical rigour. For example, instead of being limited to saying: It took me three hours to walk from the centre of London to Greenwich, a distance of six kilometres, therefore my average speed was:

$$\bar{v} = \frac{6 \text{ km}}{3 \text{ hours}} = 2 \text{ km/hour} \tag{1.14}$$

It was now possible to look more closely at speed and in particular the instantaneous speed could be defined as:

$$v = \lim_{\Delta t \to 0} \frac{\Delta s}{\Delta t} \tag{1.15}$$

which was expressed succinctly as a differential:

$$v = \frac{ds}{dt} \tag{1.16}$$

It was this step in understanding that then led to the development of what is now called 'Classical Mechanics'. Under this more mathematically precise framework, the acceleration:

$$a = \frac{\Delta v}{\Delta t} \quad \text{becomes} \quad a = \frac{dv}{dt} \tag{1.17}$$

but v itself is a differential, as given by equation (1.16), therefore:

$$a = \frac{d}{dt}\frac{ds}{dt} = \frac{d^2 s}{dt^2} \tag{1.18}$$

Acceleration as a Differential

Contrary to what had been thought at the time, Isaac Newton discovered that a particle always continued its motion along a straight line with a constant speed *unless* acted on by a 'force'—an observation which came to be known as Newton's First Law. Newton also summarised the obvious converse: If a particle is subjected to a force then it experiences a change in its motion. But the major breakthrough Newton made was to quantify this mathematically in what has how become known as Newton's Second Law: The rate of change of momentum of a body is proportional to the force acting upon it, which in the new mathematics would be written as:

$$F = \frac{dp}{dt} = \frac{d(mv)}{dt} \tag{1.19}$$

Newton's Second Law

Excluding space rockets the vast majority of everyday objects have a mass which remains constant with time, hence in the above equation the mass m can be brought outside of the differential, i.e.

$$F = m\frac{dv}{dt} \tag{1.20}$$

which can, of course, be written in the more familiar form:

$$F = ma \tag{1.21}$$

Again, the falling cannonball is a good example of this: The cannonball accelerates downwards at a rate g due to an attractive force between its mass m and the Earth, this force is commonly referred to as its 'weight'.

It was clear that moving objects possessed 'energy'—the ability to do work, and that this energy could be transferred from one type to another. For example, if you have a cannonball, initially at rest, at the top of a hill, then clearly it has no energy associated with its motion. However, if the cannonball were given a small push, just enough to start

3

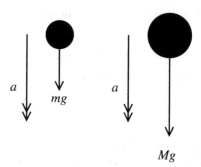

Fig. 1.1 Cannonballs (like all other objects) fall downwards because they feel a force—their 'weight'. The weight is proportional to their mass and hence they accelerate at the same rate

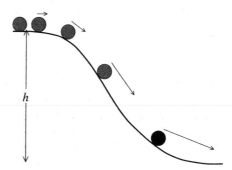

Fig. 1.2   The transfer of potential to kinetic energy

it rolling, it would roll down the hill and gain in speed, see Fig. 1.2. The cannonball now has motional energy, and the faster it goes, the more sense it makes to talk of an increase in this energy. This type of energy became known as 'kinetic energy'. But where had this energy come from? Had it materialised out of thin air? Although a pleasant thought, observations and further ponderings suggested that energy was merely changed from one type to another. For example, the kinetic energy of the cannonball rolling down the hill could be converted to sound energy if it collided with a brick wall. The kinetic energy of water could be converted into rotational energy by a waterwheel, the chemical energy in coal and wood could be converted to heat and light by burning and so on. Thus the concept of 'potential energy' was created to describe objects and particles in situations *Potential* where they could gain kinetic energy if allowed to do so, i.e. push the cannonball over the *Energy* edge, open the sluice gate to allow the water to flow, etc.

The concept of 'work' was introduced to describe the amount of effort expended in achieving an objective. Consider carrying (or pushing) the cannonball to the top of the hill in the first place. Work has to be done against the force of gravity, i.e. against the weight of the cannonball. The heavier (mg) the ball, the harder it is, i.e. the more work that has to be done, in addition the higher 'h' the hill (see Fig. 1.2) the larger the required effort, thus the work done is quantified mathematically as:

Concept of Work

$$\text{work done} = \text{force} \times \text{distance moved in direction of force} \qquad (1.22)$$

which in this case would be summarised mathematically as:

$$w = mgh \qquad (1.23)$$

This work is stored within the cannonball-hill system, and provided the ball is nudged over the edge, it can be regained as kinetic energy. Hence, the stored work is assigned with the potential energy concept, i.e. labelling it V, then the potential energy associated with the gravitational field is given by $V = mgh$.

By the same route the kinetic energy, call it T, of a particle is equal to the amount of energy expended by a force in accelerating the particle from rest to its final velocity v, i.e.

$$T = \text{force} \times \text{distance moved in direction of force} \qquad (1.24)$$

By Newton's Second Law the force is equal to the rate of change of momentum, as in equation (1.19). Therefore if the force is applied over a distance s, then:

$$T = \frac{\mathrm{d}(mv)}{\mathrm{d}t} s \qquad (1.25)$$

Again excluding space rockets, and writing within the new calculus, then:

$$T = \int_0^s m\frac{\mathrm{d}v}{\mathrm{d}t}\ \mathrm{d}s \qquad (1.26)$$

Recalling that the speed is the rate of change of position, i.e. $v = \mathrm{d}s/\mathrm{d}t$, then:

$$T = m\int_0^s \mathrm{d}v\frac{\mathrm{d}s}{\mathrm{d}t} = m\int_0^v v\mathrm{d}v = \frac{1}{2}mv^2 \qquad (1.27)$$

Definition of Kinetic Energy

i.e. the kinetic energy of a particle is proportional to the square of its speed, as illustrated in Fig. 1.3.

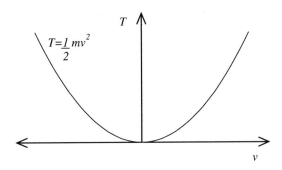

Fig. 1.3 The kinetic energy T of a particle as a function of its speed v

Clearly, an important quantity would be the total energy of a system, which within the framework developed here would be a sum of the kinetic and potential energies $T + V$.

Hamilton generalised this further to describe the total energy of a group of particles of different masses m_i and speeds v_i, say. This entity is now called the 'Hamiltonian':

$$H = T + V = \sum_{i=1}^{N} \left(\frac{1}{2} m_i v_i^2 + V_i \right) \tag{1.28}$$

where V_i is the potential energy of the ith particle.

Such a collection of particles could be the atoms in a gas or a solid, where the potential energies in that case would be electrostatic due to the Coulomb force between the electric charges. Or, on a more everyday basis, in Nineteenth Century Britain a popular pastime was the village football game. In this game a goal was placed at each end of the village and a football (made of an inflated pig's bladder) was kicked, carried and thrown between two teams of players. The rules were few and the field often muddy and sloped, as depicted in Fig. 1.4.

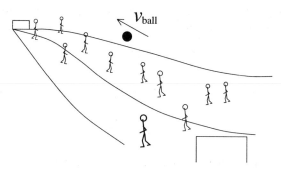

Fig. 1.4 The village football match

If the speed and mass of the ith player were v_i and m_i respectively, given that the potential energy consists entirely of gravitational potential energy, which is equal to the weight of each player times their height above the lowest point on the field, then the Hamiltonian representing the football match would be:

Hamiltonian for a Football Match

$$H = T + V = \sum_{i=1}^{N} \left(\frac{1}{2} m_i v_i^2 + \frac{1}{2} m_{\text{ball}} v_{\text{ball}}^2 + m_i g h_i + m_{\text{ball}} g h_{\text{ball}} \right) \tag{1.29}$$

All physical systems strive to minimise their energy, and after 3 hours of running up and down and kicking lumps out of each other, the ball would come to rest and all the players would collapse onto the floor (minimising their kinetic energy) at the bottom of the hill (minimising their potential energy).

1.2 WAVES AND QUANTUM MECHANICS

At the beginning of the Twentieth Century, experiments such as the photoelectric effect, and the theoretical explanation of black-body radiation, indicated that electromagnetic

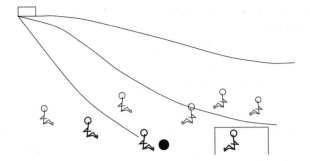

Fig. 1.5 The end of the match

radiation hitherto thought of as waves, actually displayed some properties which were 'particle-like', i.e. light was quantised in small packets of energy which are now called 'photons'. In addition, experiments such as electron diffraction through a crystal suggested that electrons could behave like waves. Thus was born the concept of 'wave–particle duality'.

De Broglie summarised this wave–particle duality by stating that a particle of momentum p has an associated wave of wavelength λ given by the following:

$$\lambda = \frac{h}{p} \qquad (1.30)$$

De Broglie's Relation

where the constant h was derived by Planck in explaining black-body radiation.

Thus, consider an electron at a position z and moving along a straight line in a vacuum and away from the influence of any electromagnetic, or indeed gravitational, potentials. Then under the auspices of wave–particle duality this particle could be described by a *state function* which is of the form of a wave, i.e.

$$\psi = e^{i(kz - \omega t)} \qquad (1.31)$$

Equation of a Wave

where t is the time, ω the angular frequency and the wave vector is given by:

$$k = \frac{2\pi}{\lambda} \qquad (1.32)$$

The question then was: 'What do these waves represent?' and this occupied the greatest scientific minds for the best part of two decades. Indeed by comparison with the continuity equation it became apparent that the modulus squared of the state or 'wave function' ($\psi^*\psi$) should be interpreted as the probability of a particle (or wave) being at a certain position in space at a particular time. And it was the manipulations of the wave functions, carried out in order to deduce more familiar physical observables, such as energy and momentum, that became known as the theory of 'wave' or 'quantum' mechanics.

In particular, the quantum mechanical momentum was deduced to be a linear operator acting upon the wave function ψ, with the momentum p arising as an eigenvalue, i.e.

$$-i\hbar\frac{\partial \psi}{\partial z} = p\psi \qquad (1.33)$$

Quantum Momentum Operator

which when acting upon the wave function of the electron in the vacuum, given by equation (1.31), would give:

$$-i\hbar\frac{\partial}{\partial z}e^{i(kz-\omega t)} = pe^{i(kz-\omega t)} \tag{1.34}$$

the nature of the eigenvalue equation can then be seen:

$$-i\hbar \times ike^{i(kz-\omega t)} = pe^{i(kz-\omega t)} \tag{1.35}$$

i.e. the linear momentum $p = \hbar k$, which, not surprisingly, can be manipulated ($p = \hbar k = (h/2\pi)(2\pi/\lambda) = h/\lambda$) to reproduce de Broglie's relationship in equation (1.30).

Quantum
Momentum
Operator

Such an approach could be easily generalised to a particle or wave moving in three dimensions:

$$-i\hbar\nabla\psi = \mathbf{p}\psi \tag{1.36}$$

where the spatial derivative ∇ is given by:

$$\nabla = \frac{\partial}{\partial x}\mathbf{i} + \frac{\partial}{\partial y}\hat{\mathbf{j}} + \frac{\partial}{\partial z}\hat{\mathbf{k}} \tag{1.37}$$

If this acted upon a more general wave function for an electron in a vacuum, which was of the form $e^{i(\mathbf{k}\cdot\mathbf{r}-\omega t)}$ then:

$$-i\hbar\nabla e^{i(\mathbf{k}\cdot\mathbf{r}-\omega t)} = \mathbf{p}e^{i(\mathbf{k}\cdot\mathbf{r}-\omega t)} \tag{1.38}$$

and therefore

$$-i\hbar\left(\frac{\partial}{\partial x}\mathbf{i} + \frac{\partial}{\partial y}\hat{\mathbf{j}} + \frac{\partial}{\partial z}\hat{\mathbf{k}}\right)e^{i(k_x x+k_y y+k_z z-\omega t)} = \mathbf{p}e^{i(\mathbf{k}\cdot\mathbf{r}-\omega t)} \tag{1.39}$$

$$\therefore -i\hbar\left(ik_x\mathbf{i} + ik_y\hat{\mathbf{j}} + ik_z\hat{\mathbf{k}}\right)e^{i(k_x x+k_y y+k_z z-\omega t)} = \mathbf{p}e^{i(\mathbf{k}\cdot\mathbf{r}-\omega t)} \tag{1.40}$$

Thus the eigenvalue:

$$\mathbf{p} = \hbar\left(k_x\mathbf{i} + k_y\hat{\mathbf{j}} + k_z\hat{\mathbf{k}}\right) = \hbar\mathbf{k} \tag{1.41}$$

which is again consistent with preconceived notions from experience with classical mechanics.

Classical mechanics gave the kinetic energy of a particle of mass m, travelling with a speed v, as:

$$T = \frac{1}{2}mv^2 = \frac{(mv)^2}{2m} = \frac{p^2}{2m} \tag{1.42}$$

Following the same principle as with the linear momentum, it may be hoped that the quantum mechanical analogy for the kinetic energy can also be represented by an eigenvalue equation. This time the operator, $\mathcal{T}$ say, would have to consist of the linear momentum operator $\mathcal{P}$ acting twice and then dividing by twice the mass, [1] i.e.

$$\mathcal{T} = \frac{1}{2m}\mathcal{P}\mathcal{P} \tag{1.43}$$

---

[1] This order of the operators composing $\mathcal{T}$, assumes that the mass m is constant. This is not true under certain models of solid state systems and a different kinetic energy operator must be derived, see Harrison, *Quantum Wells, Wires and Dots*, (Wiley, Chichester, 1999).

with the kinetic energy arising as an eigenvalue T when acting upon the wave function:

$$\mathcal{T}\psi = T\psi \tag{1.44}$$

Using the one-dimensional form for the linear momentum operator $\mathcal{P}$, as in equation (1.33), then:

$$\mathcal{T} = \frac{1}{2m}\left(-i\hbar\frac{\partial}{\partial z}\right)\left(-i\hbar\frac{\partial}{\partial z}\right) = -\frac{\hbar^2}{2m}\frac{\partial^2}{\partial z^2} \tag{1.45}$$

Therefore:

$$-\frac{\hbar^2}{2m}\frac{\partial^2}{\partial z^2}\psi = T\psi \tag{1.46}$$ *Kinetic Energy Operator*

When acting upon the electron vacuum wave function given in equation (1.31), then:

$$-\frac{\hbar^2}{2m}\frac{\partial^2}{\partial z^2}e^{i(kz-\omega t)} = Te^{i(kz-\omega t)} \tag{1.47}$$

$$\therefore -\frac{\hbar^2}{2m}\left(i^2k^2\right)e^{i(kz-\omega t)} = Te^{i(kz-\omega t)} \tag{1.48}$$

Thus the kinetic energy eigenvalue is given by:

$$T = \frac{\hbar^2 k^2}{2m} \tag{1.49}$$

For an electron in a vacuum away from the influence of electromagnetic fields, then the total energy E is just the kinetic energy T. Thus the dispersion or energy versus momentum (which is proportional to the wave vector k) curves are parabolic, see Fig. 1.6.

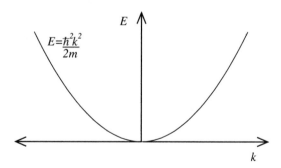

Fig. 1.6 The energy versus wave vector (proportional to momentum) curve for an electron in a vacuum

This is the same result as for classical free particles, derived earlier and illustrated in Fig. 1.3 (the momentum is proportional to the speed for a particle of constant mass).

Again, this operator can be generalised to three dimensions by utilising the three-dimensional form for the linear momentum operator given in equation (1.36), i.e.

$$\mathcal{T} = \frac{1}{2m}\left(-i\hbar\nabla\right)\left(-i\hbar\nabla\right) \tag{1.50}$$

which, given the form of ∇ in equation (1.37), becomes:

$$-\frac{\hbar^2}{2m}\left(\frac{\partial^2}{\partial x^2} + \frac{\partial^2}{\partial y^2} + \frac{\partial^2}{\partial z^2}\right) \tag{1.51}$$

and when acting upon the electron vacuum wave function gives:

$$T = \frac{\hbar^2}{2m}\mathbf{k}\bullet\mathbf{k} = \frac{\hbar^2 k^2}{2m} \tag{1.52}$$

Look again at the form for the kinetic energy operator in either equation (1.46) or equation (1.51). Clearly all solutions for the wave function and its first derivative must be continuous, otherwise the differential operator will yield infinite values. This is an important result which is worth stating mathematically (in one-dimension):

Wave Function Boundary Conditions

$$\psi(z) \quad \text{and} \quad \frac{\partial\psi}{\partial z} \quad \text{are continuous} \tag{1.53}$$

Now consider the possibility that the electron has some potential energy, perhaps due to the presence of an electric field or gravitational field, as in Fig. 1.7.

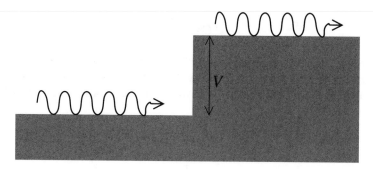

Fig. 1.7   An electron in the presence of a potential

The Hamiltonian of the electron would then be composed of two parts, the familiar kinetic energy component, T, as derived above, and clearly the potential energy component V which is just added to form the total energy:

The Total Energy

$$H = T + V \tag{1.54}$$

In quantum mechanics, these components are represented by eigenvalues of operators, the former being a second-order differential operator as derived above. Given that the familiar electrostatic and gravitational potentials are not functions of energy or momentum, then the quantum mechanical potential operator $\mathcal{V}$ say, is just a straightforward scalar multiplier with the potential energy V arising as an eigenvalue:

Potential Energy Operator

$$\mathcal{V}\psi = V\psi \tag{1.55}$$

clearly ψ is an eigenvector of $\mathcal{V}$. Thus using the one-dimensional form for the kinetic energy operator, the Hamiltonian operator in quantum mechanics would be given by:

$$\mathcal{H} = -\frac{\hbar^2}{2m}\frac{\partial^2}{\partial z^2} + V \qquad (1.56)$$

Hamiltonian (Total Energy) Operator

or in three dimensions:

$$\mathcal{H} = -\frac{\hbar^2}{2m}\nabla^2 + V \qquad (1.57)$$

When coupled with the wave function ψ, the resulting eigenvalue equation, specifying the total energy E, i.e.

$$\mathcal{H}\psi = E\psi \qquad (1.58)$$

Time-Independent Schrödinger Equation

is known as the time-independent Schrödinger equation. It has this name because the Hamiltonian $\mathcal{H}$ doesn't depend on time, but the wave function ψ does have a time dependency.

Consider now operating on the wave function with the linear operator $i\hbar\partial/\partial t$, i.e.

$$i\hbar\frac{\partial}{\partial t}e^{i(kz-\omega t)} = i\hbar(-i\omega)e^{i(kz-\omega t)} \qquad (1.59)$$

i.e.

$$i\hbar\frac{\partial}{\partial t}\psi = \hbar\omega\psi \qquad (1.60)$$

Clearly this eigenvalue $\hbar\omega$ is also the total energy but in a form usually associated with waves, e.g. a photon. These two operations on the wave function represent the two complementary descriptions associated with *wave–particle duality* and the equation

$$i\hbar\frac{\partial}{\partial t}\psi = E\psi \qquad (1.61)$$

Time-Dependent Schrödinger Equation

is known as the time-dependent Schrödinger equation .

1.3 ANALYTICAL SOLUTIONS TO SCHRÖDINGER'S EQUATION

Different physical systems are represented by different *potentials*. So whereas the kinetic energy operator (nearly) always stays the same, the quantum mechanics is summarised by specifying the potential.

Some potentials have analytical solutions, for example:

- Constant potentials, i.e. $V = $ constant: If $E > V$ then in one dimension the solutions are a linear combination of waves of the type met above:

$$\psi = Ae^{ikz} + Be^{-ikz} \qquad (1.62)$$

This can be verified by substituting into the corresponding Schrödinger equation formed from the Hamiltonian in equation (1.56):

$$\left(-\frac{\hbar^2}{2m}\frac{\partial^2}{\partial z^2} + V\right)\left(Ae^{ikz} + Be^{-ikz}\right) = E\left(Ae^{ikz} + Be^{-ikz}\right) \qquad (1.63)$$

$$\therefore -\frac{\hbar^2}{2m}\left[A(ik)^2 e^{ikz} + B(-ik)^2 e^{-ikz}\right] = (E - V)\left(Ae^{ikz} + Be^{-ikz}\right) \quad (1.64)$$

$$\therefore \frac{\hbar^2 k^2}{2m}\left(Ae^{ikz} + Be^{-ikz}\right) = (E - V)\left(Ae^{ikz} + Be^{-ikz}\right) \quad (1.65)$$

equating coefficients, then:

$$\frac{\hbar^2 k^2}{2m} = (E - V) \qquad \therefore k = \frac{\sqrt{2m(E - V)}}{\hbar} \quad (1.66)$$

If $E < V$ then equation (1.66) would imply that:

$$k = \frac{\sqrt{-2m(V - E)}}{\hbar} \quad \text{in which case} \quad k = i\frac{\sqrt{2m(V - E)}}{\hbar} \quad (1.67)$$

writing k as $i\kappa$, then the original wave function in equation (1.62) would become:

$$\psi = Ae^{i(i\kappa)z} + Be^{-i(i\kappa)z} = Ae^{-\kappa z} + Be^{\kappa z} \quad (1.68)$$

i.e. a linear combination of growing and decaying exponentials. These are the familiar solutions[2] to the so-called infinitely deep and finite 'quantum wells'.

Thus considering the Schrödinger equation as just a mathematical construct, allows solutions for the wave function ψ to be deduced. However, this is not the whole story. As with the solution of any differential equation it is necessary to impose some initial (or boundary) conditions to describe the solution fully, i.e. to be able to specify the unknown constants A and B. In quantum mechanics these arise from a consideration of the *physics*—it has already been mentioned that $\psi^*\psi$ should be interpreted as the probability density, thus as the particle must exist somewhere, the integral over all space of $\psi^*\psi$ must be unity. In a one-dimensional system this would be expressed mathematically as:

Probability Normalisation Condition

$$\int_{-\infty}^{+\infty} \psi^*(z)\psi(z) \ \mathrm{d}z = 1 \quad (1.69)$$

For this to be the case then clearly:

Standard Boundary Conditions

$$\psi - \!\!\rightarrow 0 \quad \text{as} \quad z \longrightarrow \infty \quad (1.70)$$

and further:

$$\frac{\partial \psi}{\partial z} - \!\!\rightarrow 0 \quad \text{as} \quad z \longrightarrow \infty \quad (1.71)$$

The latter two expressions are known as the 'standard boundary conditions' and are a valuable tool in constructing quantum solutions.

---

[2]See any introductory text on modern physics or quantum mechanics.

- Parabolic potentials, i.e. $V \propto z^2$: Again in one dimension with the spatial coordinate z, then if the potential is proportional to the square of the distance from some fixed point (origin), the Schrödinger equation is of the form:

$$\left(-\frac{\hbar^2}{2m}\frac{\partial^2}{\partial z^2} + C\frac{z^2}{2}\right)\psi = E\psi \qquad (1.72)$$

where C is some constant. This is the quantum mechanical analogy to the classical harmonic oscillator, in which if a mass m was displaced from the equilibrium position by an amount z, then it would experience a restoring force proportional to the gradient in the potential and in the opposite direction, i.e.

$$\text{restoring force} = -\frac{\partial V}{\partial z} = -Cz \qquad (1.73)$$

Hence the mass would undergo simple harmonic motion with oscillations of angular frequency:

$$\omega = \sqrt{\frac{C}{m}} \qquad (1.74)$$

The behaviour of a quantum particle within such a potential is quite different and in contrast to the classical system, the eigenvalues of the Schrödinger equation form a ladder of equally spaced energies:

Quantised States of Harmonic Oscillator

$$E_n = \left(n - \frac{1}{2}\right)\hbar\omega, \quad \text{where} \quad n = 1, 2, 3, \ldots \qquad (1.75)$$

where the convention that the lowest energy state is labelled with $n = 1$ has been applied, and

$$\hbar\omega = \hbar\sqrt{\frac{C}{m}} \qquad (1.76)$$

Such potentials are important in solid state physics where, for example, the motion of atoms bound to neighbouring atoms by electrostatic chemical bonds are similar to mass-spring oscillators.

- Sloped potentials: Consider the general linear sloped potential:

$$V(z) = Fz \qquad (1.77)$$

where F is some constant of proportionality; this would be the exact form if the vertex in Fig. 1.8 was at the origin.

Then the complete Schrödinger equation would be:

$$-\frac{\hbar^2}{2m}\frac{\partial^2\psi}{\partial z^2} + Fz\psi = E\psi \qquad (1.78)$$

Sloped Potentials

$$\therefore \frac{\partial^2\psi}{\partial z^2} - \frac{2m}{\hbar^2}\left(Fz - E\right)\psi = 0 \qquad (1.79)$$

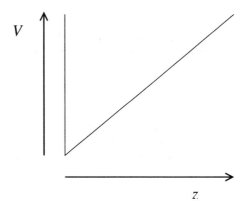

Fig. 1.8   A general sloped $(V(z) = Fz)$ potential

Then by making the substitutions:

$$\alpha = \frac{2m}{\hbar^2}(-E) \quad \text{and} \quad \beta = \frac{2m}{\hbar^2}F \tag{1.80}$$

the Schrödinger equation becomes:

$$\frac{\partial^2 \psi}{\partial z^2} - (\alpha + \beta z)\,\psi = 0 \tag{1.81}$$

with the further substitution $z' = (\alpha + \beta z)/\gamma$, then:

$$\frac{\beta^2}{\gamma^2}\frac{\partial^2 \psi}{\partial z'^2} - \gamma z'\psi = 0 \tag{1.82}$$

If γ^3 is chosen to be equal to β^2, then equation (1.82) becomes:

*Airy's
Equation*

$$\frac{\partial^2 \psi}{\partial z'^2} - z'\psi = 0 \tag{1.83}$$

which is Airy's equation[3] and has analytical solutions which are linear combinations of the Airy functions $\mathrm{Ai}(z')$ and $\mathrm{Bi}(z')$.

This type of analytical solution to Schrödinger's equation is important in semiconductor device physics, because an electric field applied to a semiconductor produces sloped potentials of this form. The charge carriers, whether they be electrons or holes 'roll' down this potential and gain kinetic energy, thus constituting an electric current. The triangular potential shown in Fig. 1.8 resembles that produced at the doped heterojunction between two dissimilar semiconductors in a High-Electron-Mobility-Transistor or HEMT.

---

[3]See Abramowitz and Stegun, *Handbook of Mathematical Functions*, (Dover, New York, 1965), p. 446.

- Pöschl-Teller potential holes: This one-dimensional potential is given by:

$$V(z) = -\frac{\hbar^2}{2m}\alpha^2\frac{\lambda(\lambda-1)}{\cosh^2\alpha z}$$

(1.84)   *Pöschl-Teller Potentials*

where α and λ are two parameters which adjust the width and depth of the hole. As can be seen from equation (1.84), α has dimensions of (length)$^{-1}$ and λ is dimensionless. An example, with $\alpha = 0.05$ and $\lambda = 2.0$, is shown in Fig. 1.9.

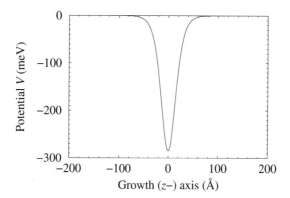

Fig. 1.9 Pöschl-Teller potential hole; $\alpha = 0.05$ and $\lambda = 2.0$. Reproduced with permission from John Wiley & Sons Ltd

The eigenvalues of the resulting Schrödinger equation are given by:

$$E_n = -\frac{\hbar^2\alpha^2}{2m}(\lambda - 1 - n)^2 \quad n = 1, 2, 3, \ldots$$

(1.85)

The potential profile of a Pöschl-Teller potential hole is often used to approximate that produced by the diffusion of material species at double heterojunctions (or quantum wells) found increasingly in modern opto-electronic devices[4].

There are a range of potentials whose corresponding Schrödinger's equations do not have analytical solutions, for example, $V(z) \propto z^4$, or higher orders, potentials produced by the diffusion of material species at semiconductor heterojunctions, potentials produced by distributions of electrons and holes in semiconductor devices or by the atoms in a crystalline or non-crystalline material, or indeed the potential within an atom produced by the nuclear charge and the surrounding electron clouds. In fact, analytical solutions to Schrödinger's equation are quite rare and occur only for simple highly-symmetric potentials.

---

[4]See Kelly, *Low-Dimensional Semiconductors*, (Clarendon Press, Oxford, 1995).

1.4 THE SHOOTING METHOD

There is therefore an impetus for alternative methods of solution of Schrödinger's equation which do not rely upon lengthy algebraic manipulation of complicated esoteric mathematical functions *for each different form of the potential.* The rapid development of calculating machines and computers has provided such an opportunity to tackle potentials that would otherwise be avoided.

In this chapter attention will be focussed on *one* of the methods of solving Schrödinger's equation which is just a second-order differential equation. There are other techniques of solving differential equations[5], some of which are more sophisticated, and some of which may be even more suited to particular potentials, however, the method illustrated here is a good place to start. It is quick to derive and just as important, it is straightforward to implement—which in this context means 'construct a computer program which gives numbers out'.

Consider the one-dimensional time-independent Schrödinger equation, some of the analytical solutions of which have been extensively studied in the previous section:

$$-\frac{\hbar^2}{2m}\frac{\partial^2}{\partial z^2}\psi(z) + V(z)\psi(z) = E\psi(z) \tag{1.86}$$

where the one-dimensional potential $V(z)$ will remain undefined and again $\psi(z)$ is the wave function representing the particle of interest. This, of course, can be written as follows:

$$-\frac{\hbar^2}{2m}\frac{\partial^2}{\partial z^2}\psi(z) + [V(z) - E]\,\psi(z) = 0 \tag{1.87}$$

The problem now is to find a numerical method for the solution of both the energy eigenvalues E and the eigenfunctions $\psi(z)$ for *any* $V(z)$.

With this aim, consider expanding the second-order derivative in terms of finite differences. For example, as in Fig. 1.10, the first derivative of any function $f(z)$ is defined as:

$$\lim_{\Delta z \to 0} \frac{\Delta f}{\Delta z} = \frac{\mathrm{d}f}{\mathrm{d}z} \tag{1.88}$$

It suits the purpose here to retain the approximate form, i.e.

Approximation to First Derivative

$$\frac{\mathrm{d}f}{\mathrm{d}z} \approx \frac{\Delta f}{\Delta z} = \frac{f(z + \delta z) - f(z - \delta z)}{2\delta z} \tag{1.89}$$

The first derivative, say $f'(z)$, is just another function, and its derivative can be found in the same way, i.e.

$$\frac{\mathrm{d}f'}{\mathrm{d}z} \approx \frac{f'(z + \delta z) - f'(z - \delta z)}{2\delta z} \tag{1.90}$$

Or, put more mathematically, the second derivative follows as:

$$\frac{\mathrm{d}^2 f}{\mathrm{d}z^2} \approx \frac{\left.\frac{\mathrm{d}f}{\mathrm{d}z}\right|_{z+\delta z} - \left.\frac{\mathrm{d}f}{\mathrm{d}z}\right|_{z-\delta z}}{2\delta z} \tag{1.91}$$

---

[5] See, for example, W. H. Press *et al.*, *Numerical Recipes in C: the art of scientific computing* (University Press, Cambridge, 1992).

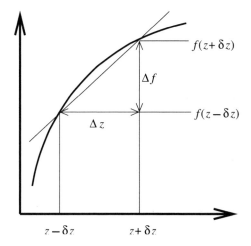

Fig. 1.10   The first derivative of a function

By using the finite difference forms in equation (1.89) for the first derivatives, then:

$$\frac{d^2 f}{dz^2} \approx \frac{\left[\frac{f(z+2\delta z)-f(z)}{2\delta z}\right] - \left[\frac{f(z)-f(z-2\delta z)}{2\delta z}\right]}{2\delta z} \tag{1.92}$$

$$\therefore \frac{d^2 f}{dz^2} \approx \frac{f(z + 2\delta z) - 2f(z) + f(z - 2\delta z)}{(2\delta z)^2} \tag{1.93}$$

As δz is an, as yet, undefined small step length along the z-axis, and as it only appears in equation (1.93) with the factor 2, then this finite difference representation of the second derivative can be simplified slightly by substituting δz for $2\delta z$, i.e.

$$\frac{d^2 f}{dz^2} \approx \frac{f(z + \delta z) - 2f(z) + f(z - \delta z)}{(\delta z)^2} \tag{1.94}$$

Approximation to Second Derivative

Thus, whereas the first derivative of a function can be approximated by calculating the slope between two closely spaced points, the *second* derivative can be approximated by knowing the value of a function at three points, see Fig. 1.11.

Using this form for the second derivative in the original Schrödinger equation (equation (1.87)) and taking the step length δz as sufficiently small that the approximation is good, i.e. the '$\approx$' can be dropped in favour of an '=', then:

$$-\frac{\hbar^2}{2m^*}\left[\frac{\psi(z + \delta z) - 2\psi(z) + \psi(z - \delta z)}{(\delta z)^2}\right] + [V(z) - E]\,\psi(z) = 0 \tag{1.95}$$

$$\therefore \psi(z + \delta z) - 2\psi(z) + \psi(z - \delta z) = \frac{2m^*}{\hbar^2}(\delta z)^2\,[V(z) - E]\,\psi(z) \tag{1.96}$$

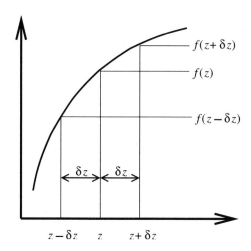

Fig. 1.11   The second derivative of a function

which can finally be written as:

*The Shooting
Equation*

$$\psi(z + \delta z) = \left[\frac{2m^*}{\hbar^2} (\delta z)^2 \left(V(z) - E \right) + 2 \right] \psi(z) - \psi(z - \delta z) \qquad (1.97)$$

Equation (1.97) implies that if the wave function is known at the two points $(z - \delta z)$ and z, then the value of the wave function at $(z + \delta z)$ can be calculated for any given energy E. This iterative equation forms the basis of a standard method of solving differential equations numerically, and is known as the 'shooting method'.

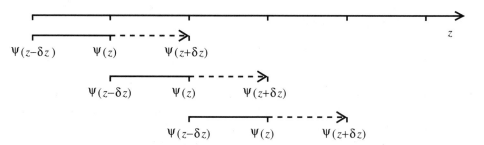

Fig. 1.12   The shooting method

Using two known values of the wave function $\psi(z - \delta z)$ and $\psi(z)$, a third value, i.e. $\psi(z + \delta z)$, can be predicted, see third line from bottom of Fig. 1.12. Using this new point $\psi(z + \delta z)$, together with $\psi(z)$ and by making the transformation $z + \delta z \to z$, a fourth point, $\psi(z + 2\delta z)$, can be calculated (second line from bottom of Fig. 1.12), and so on. Hence a *mathematical* solution to Schrödinger's equation can be deduced for any particular energy. The *physically correct* solutions occur for those energies E where the solutions satisfy the standard boundary conditions, i.e.

Standard Boundary Conditions

$$\psi(z) \to 0 \quad \text{and} \quad \frac{\partial}{\partial z}\psi(z) \to 0, \quad \text{as} \quad z \to \pm\infty \qquad (1.98)$$

The mathematical solutions to Schrödinger's equation ψ are really functions of both position z and energy E, i.e. $\psi = \psi(z, E)$, and the physical solutions are selected according to the criteria that they satisfy the standard boundary conditions in equation (1.98). The selection procedure can be summarised mathematically by defining the energy eigenvalues as the solutions to the equation:

$$\psi(\infty, E) = 0 \qquad (1.99)$$

1.5 INITIAL CONDITIONS

An iterative technique for deducing mathematical, and then selecting physical, solutions to Schrödinger's equation has been formulated *provided* that two initial wave function points are known. To deduce the latter it is necessary to consider the form of the potential in more detail, and in particular consider the special (though common) case of a potential symmetric about some origin ($z = 0$), i.e.

A Symmetric Potential

$$V(z) = V(-z) \qquad (1.100)$$

Note, the parabolic and Pöschl-Teller potentials discussed earlier are of this form. In general, solutions to Schrödinger's equation (i.e. wave functions) take the same symmetry as the Hamiltonian defining the system, which means that for the symmetric potential above, the wave functions are either symmetric, i.e.

$$\psi(z) = \psi(-z) \qquad (1.101)$$

or anti-symmetric:

$$\psi(z) = -\psi(-z) \qquad (1.102)$$

At the origin, the latter can only be true if $\psi(0) = 0$. Furthermore, the value of the wave function at a small step along the z–axis, i.e. $\psi(\delta z)$ is not important (as long as it is non-zero) because the energy eigenvalues of the Schrödinger equation are not altered when the wave function is scaled by a constant factor. To see this, consider the Schrödinger equation:

$$-\frac{\hbar^2}{2m}\frac{\partial^2}{\partial z^2}\psi(z) + V(z)\psi(z) = E\psi(z) \qquad (1.103)$$

Now multiply $\psi(z)$ by some constant ξ, then:

$$-\frac{\hbar^2}{2m}\frac{\partial^2}{\partial z^2}\left[\xi\psi(z)\right] + V(z)\left[\xi\psi(z)\right] = E\left[\xi\psi(z)\right] \qquad (1.104)$$

Now the differential is a linear operator, and the potential and energy eigenvalue just multipliers, hence the constant ξ can be brought out to the front:

$$-\xi\frac{\hbar^2}{2m}\frac{\partial^2}{\partial z^2}\psi(z) + \xi V(z)\psi(z) = \xi E\psi(z) \qquad (1.105)$$

and divided through to leave the eigenvalue E unchanged. Thus the starting values for the iterative procedure for anti-symmetric (often referred to as odd-parity) solutions of symmetric potentials can be taken as:

Odd-Parity
Initial
Conditions

$$\psi(0) = 0 \quad \psi(\delta z) = 1 \tag{1.106}$$

Now consider the starting conditions for the symmetric (even-parity) solutions. The symmetry of the solution implies that at a small step δz either side of the origin, the wave function is equal, i.e. $\psi(-\delta z) = \psi(\delta z)$. Choosing $z = 0$ in the shooting equation (equation (1.97)) then:

$$\psi(\delta z) = \left[\frac{2m^*}{\hbar^2}(\delta z)^2 \left(V(0) - E \right) + 2 \right] \psi(0) - \psi(-\delta z) \tag{1.107}$$

Now, of course, $\psi(\delta z) = \psi(-\delta z)$ therefore:

$$2\psi(\delta z) = \left[\frac{2m^*}{\hbar^2}(\delta z)^2 \left(V(0) - E \right) + 2 \right] \psi(0) \tag{1.108}$$

As discussed above, as the wave function can be scaled by a constant factor without changing the energy eigenvalues, then choose $\psi(0) = 1$, the second starting value $\psi(\delta z)$ then follows from equation (1.108), and summarising:

Even-Parity
Initial
Conditions

$$\psi(0) = 1 \quad \psi(\delta z) = \frac{m^*}{\hbar^2}(\delta z)^2 \left[V(0) - E \right] + 1 \tag{1.109}$$

1.6   COMPUTATIONAL IMPLEMENTATION

So the mathematical framework has now been derived for a numerical solution to the one-dimensional Schrödinger equation for symmetric potentials. However, as the aim is to be able to implement this on a computer and hence produce numerical answers for potentials of interest to the reader, then consideration has to be given to what these equations mean from a computer programmer's viewpoint and how they need to be deconstructed into smaller elements which computer-based logic and decision making can understand.

In this example, such an analysis can be used to construct a relatively simple program which will work effectively enough to allow energy eigenvalues to be deduced. Additional layers of complexity can be added later to produce a more efficient and accurate algorithm.

As with any numerical method all constants have to be defined and all variables given their initial values at the beginning of the program. Following on from this an *iterative* procedure has to be implemented; in computing language this means a *loop* of some sort. An energy E is chosen and then the loop iterates (or shoots) the wave function forward along the z-axis. This is illustrated loosely in the flowchart of Fig. 1.13. The mathematics indicates that the iterative procedure should continue along the spatial axis until z reaches infinity but, of course, this cannot be achieved in reality; some 'effective infinity' must be chosen, and at this point what this is remains unknown (it will have to be deduced later by numerical experiment).

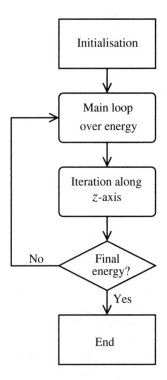

Fig. 1.13 Simplest computational implementation of the shooting method

The simple program structure illustrated with the flowchart can be broken down into smaller steps. For example, the 'Initialisation' actually means defining the numerical value of the physical constants $\hbar$ and m, as well as the computational defaults, such as the step length δz. The 'Main loop' is the part of the program which chooses the first value for the energy E and then increments it until some final value is reached and the program is halted. Within the main loop, the *mathematical* solution to the Schrödinger equation is calculated *for that particular energy E* by iterating the wave function forward from the origin towards an effective infinity. The value of the wave function at the end of this iteration can be printed out, before the energy is incremented and the next pass of the main loop made. This construct can be given more detail, as in Algorithm 1.1.

This algorithmic description is a means of breaking the mathematics into the simplest steps so that a numerical procedure can be described. The means of expression within the algorithm are intended to reflect both numerical methods and common computer programming languages—the intention being to define the technique in a *language independent format*. The statements in braces {} appearing at the end of selected lines within Algorithm 1.1 are comments intended to assist the *reader* in understanding the methodology employed by the *writer*. This is an important point, as even simple problems are often solved with quite different logic by different people, and it is important to portray the logic to anybody else who might use the solution. The arrows pointing to the left '⇐'

Algorithm 1.1 Algorithmic description of shooting method

$\hbar \Leftarrow 1.054589 \times 10^{-34}$ {Define physical constants}
$m \Leftarrow 9.109534 \times 10^{-31}$
$\delta z \Leftarrow 1 \times 10^{-10}$
for E=0 to some-upper-limit **do**
  $\psi_0 \Leftarrow 0$ {Define starting values}
  $\psi_1 \Leftarrow 1$
  for z=0 to some-effective-infinity **do**
    Calculate ψ_2
    ψ_0 takes the value of ψ_1
    ψ_1 takes the value of ψ_2
  end for
  return final value of ψ_2 {This is the value of ψ at infinity}
end for

---

can be read as 'takes the value of', for example on the first line: $\hbar$ takes the value of 1.054589×10^{-34}. The first 'for' loop creates the mechanism by which the energy E is varied from some starting value upwards. At each value of the energy E, the wave function is 'shot' forward from the starting conditions at the origin for increasing values of the spatial coordinate z. This second 'for' loop ends at the 'effective infinity' and this final value of the wave function, which is the computational equivalent of $\psi(\infty, E)$ is returned. In this simple implementation, the required energy eigenvalues, which are solutions to the equation $\psi(\infty, E) = 0$ will be found by inspection—this is perhaps best illustrated by the first example below.

Clearly there are many programming languages which would be suitable for implementing the shooting method algorithm, at this point 'C' is going to be utilised, but equally effective solutions could be found with FORTRAN, C++, BASIC, indeed virtually anything that allows for simple mathematical operations.

1.7 APPLICATION TO PARABOLIC POTENTIALS

In order to write a fully working computer code to implement the shooting technique discussed above, it is necessary to specify a potential. The first example that will be investigated will be that of an electron (of mass m_0) in a parabolic potential, one reason for this choice being the availability of the simple analytical expression for the eigenvalues (equation (1.75)) with which to compare.

Following closely the algorithmic description presented in the previous section, one particular C[6] representation for the odd-parity solutions is:

---

[6]See Appendix A for FORTRAN implementation.

*C Code
for First &
Simplest
Shooting
Implementation*

```
#define hbar 1.05459e-34
#define m 9.109534e-31
#define e_0 1.602189e-19
main()
{
float    dE=1e-3*e_0;       /* the energy increment */
float    dz=1e-10;          /* the step length delta z */
float    E;                 /* the energy E */
float    psi0,psi1,psi2;    /* psi(z-dz), psi(z), and psi(z+dz) */
float    z;                 /* the spatial coordinate */

for(E=0;E<e_0;E+=dE)
{
 psi0=0;psi1=1;                     /* set starting values */
 for(z=dz;z<100e-10;z+=dz)          /* first value of z is delta z */
 {
  psi2=(2*m*(dz/hbar)*(dz/hbar)*
       (e_0*(z/100e-10)*(z/100e-10)-E)+2)*psi1-psi0;
  psi0=psi1;
  psi1=psi2;
 }
 printf("E=%fmeV psi(infty)=%f\n",E/(1e-3*e_0),psi2);
}


}
```

It can be seen that the potential has been taken as `e_0*(z/100e-10)*(z/100e-10)`
which is mathematically equivalent to:

$$V = \left(\frac{z}{100\mathrm{\mathring{A}}}\right)^2 \mathrm{eV} \tag{1.110}$$

which is clearly parabolic (depends on z^2) and has constants of proportionality chosen
such that when $z = 100$ Å, the potential $V = 1$ eV, these are values typical of solid-state
quantum systems. Comparing with the analytical expression for the energy eigenvalues
in equation (1.75) where $\hbar\omega$ is given by equation (1.76), i.e.

$$\hbar\omega = \hbar\sqrt{\frac{C}{m}} = \hbar\sqrt{\frac{2e}{(100 \times 10^{-10})^2 m_0}} \tag{1.111}$$

which has the value of 39.039 meV.

The program increments the energy from zero (the minimum value of the potential)
upwards to 1 eV and prints the value of the wave function ψ at the effective infinity, which
has been chosen to be 100 Å.

On execution the first few lines of output appear as:

```
E=0.000000meV psi(infty)=717355352064.000000
E=1.000000meV psi(infty)=672825212928.000000
E=2.000000meV psi(infty)=630793895936.000000
E=3.000000meV psi(infty)=591129935872.000000
E=4.000000meV psi(infty)=553711042560.000000
E=5.000000meV psi(infty)=518423511040.000000
E=6.000000meV psi(infty)=485152948224.000000
E=7.000000meV psi(infty)=453795512320.000000
E=8.000000meV psi(infty)=424249917440.000000
E=9.000000meV psi(infty)=396421365760.000000
```

Further down the output it can be seen that the value of $\psi(\infty, E)$ changes from being positive to negative:

```
E=53.999994meV psi(infty)=3570301952.000000
E=54.999992meV psi(infty)=2630051072.000000
E=55.999990meV psi(infty)=1781219712.000000
E=56.999988meV psi(infty)=1017105856.000000
E=57.999986meV psi(infty)=331473376.000000
E=58.999984meV psi(infty)=-281493888.000000
E=59.999982meV psi(infty)=-827295744.000000
E=60.999980meV psi(infty)=-1311101056.000000
E=61.999979meV psi(infty)=-1737714432.000000
E=62.999977meV psi(infty)=-2111598720.000000
```

Thus it is clear that, at some energy between 57.999986 and 58.999984 meV, the wave function will tend to zero at infinity, i.e. a solution will exist. Further solutions exist: $136.00095 < E < 137.000098$ meV, $213.999890 < E < 214.999883$ meV and $291.999343 < E < 292.999336$ meV and so on. This can be illustrated more vividly by plotting $\psi(\infty, E)$ versus E as in Fig. 1.14.

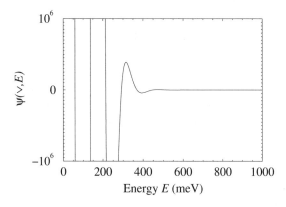

Fig. 1.14   The value of the wave function ψ at the effective infinity, versus the energy E

Each time the curve crosses the x-axis (the $y = 0$ line) a solution exists. It can be seen from Fig. 1.14 that the solutions are very regular indeed, as might be expected from the analytical solution of the parabolic potential (given earlier in equation (1.75)). Table 1.1 collates the solutions from the output of the computer program (as the energies are only known within a 1 meV range, the mid-point is chosen).

Table 1.1   Energy eigenvalues of the odd-parity solutions of a parabolic potential

n	E_n (meV)	$E_n/(n - \frac{1}{2})$
2	57.5	38.333
4	136.5	39.000
6	213.5	38.818
8	291.5	38.867
10	368.5	38.789
12	447.5	38.913

Now the analytical expression for the energy eigenvalues (equation (1.75)) gives values of $\frac{1}{2}\hbar\omega$, $1\frac{1}{2}\hbar\omega$, $2\frac{1}{2}\hbar\omega$, $3\frac{1}{2}\hbar\omega$, etc., with the second, fourth, sixth, etc. corresponding to odd-parity states. Using the numerical solutions in the second column of Table 1.1 allows a value for $\hbar\omega$ to be deduced from each solution and these are listed in the third column. It can be seen that they compare very well with the analytical solution, calculated earlier, of 39.039 meV. For example, the error is of the order of 0.1 meV which when compared to 39.039 meV is just 0.2 %. However, these solutions are not exact and furthermore they do not really improve for the higher states, which is what might be expected if the discrepancy was entirely due to the $\frac{1}{2}$ meV accuracy obtained in this quick calculation. In later sections the origin of this discrepancy will be shown and methods will be introduced which will illustrate how to improve the accuracy in the determination of the energy.

1.8   IMPROVED ENERGY EIGENVALUES

Consider the form of the function $\psi(\infty, E)$ close to a solution, as illustrated in Fig. 1.15. As the energy is incremented, the shooting method returns the value of the numerical solution to Schrödinger's equation at each energy value and the physical solutions lie between two adjacent energy points where $\psi(\infty, E)$ crosses the x-axis, as in Fig. 1.15. The actual values Y_1 and Y_2 can be used to provide an improved estimate for the energy solution by considering that the function can be approximated to a straight line between the two. Hence, considering similar triangles:

$$\frac{E - E_1}{|Y_1|} = \frac{\Delta E}{|Y_1| + |Y_2|} \tag{1.112}$$

Then a better estimate for the solution can be found by implementing this 'mid-point rule':

$$E = \frac{|Y_1|}{|Y_1| + |Y_2|}\Delta E + E_1 \tag{1.113}$$

Better Estimate of Energy

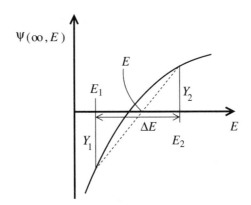

Fig. 1.15 The behaviour of the function $\psi(\infty, E)$ near a solution

or, as $E_1 = E_2 - \Delta E$:

$$E = \frac{|Y_1|}{|Y_1| + |Y_2|}\Delta E + E_2 - \Delta E \tag{1.114}$$

Programmers should always strive for neat, succinct code because, from this, larger entities may be built and the underlying structure needs to be simple and, most importantly readable. Therefore, to introduce this additional step to implement the mid-point rule, the best thing to do is to restructure the algorithm to evaluate the value of $\psi(\infty, E)$ in a subroutine or function, returning the outcome to be handled by the main program. In the algorithmic description this may appear as:

Algorithm 1.2 Algorithmic description of shooting method with mid-point rule and automatic solution selection

---

$\hbar \Leftarrow 1.054589 \times 10^{-34}$ {Define physical constants}
$m \Leftarrow 9.109534 \times 10^{-31}$
$\delta z \Leftarrow 1 \times 10^{-10}$
$Y1 \Leftarrow$ psi-at-inf(E=0) {Evaluate first value of $\psi(\infty, E)$}
for $E=\Delta E$ to some-upper-limit **do**
    $Y2 \Leftarrow$ psi-at-inf(E)
    if $Y1 \times Y2 < 0$ **then**
        Evaluate and output improved estimate using mid-point rule
    end if
    $Y1$ takes the value of $Y2$
end for

Note that, in Algorithm 1.2, the introduction of the function, given as Algorithm 1.3, allows the retention of the previous value of $\psi(\infty, E)$, hence by comparison with the current value the selection of the physical solutions from all the mathematical solutions can be made automatically. This is performed by multiplying the two values together. If the result is negative, then the function has just crossed the x-axis and a solution exists.

Algorithm 1.3 The function psi-at-inf(E)

$\psi_0 \Leftarrow 0$ {Define starting values}
$\psi_1 \Leftarrow 1$
for z=0 to some-effective-infinity **do**
     Calculate ψ_2
     ψ_0 takes the value of ψ_1
     ψ_1 takes the value of ψ_2
end for
return final value of ψ_2 {This is the value of ψ at infinity}

The C implementation of this extended algorithm is given below:

*C Code
for Improved
Shooting
Algorithm*

```
#include <math.h>

#define hbar 1.05459e-34
#define m 9.109534e-31
#define e_0 1.602189e-19
main()
{
float    psi_at_inf();      /* function psi(infinity,E) */
float    dE=1e-3*e_0;       /* the energy increment */
float    E;                 /* the energy E */
float    Y1,Y2;             /* psi(infinity,E1), psi(infinity,E2) */

Y1=psi_at_inf(0);           /* evaluate first value */

for(E=dE;E<e_0;E+=dE)       /* start loop at small value of energy */
{
 Y2=psi_at_inf(E);          /* calculate next value */
 if(Y1*Y2<0)
  printf("E=%fmeV\n",(fabs(Y1)*dE/(fabs(Y1)+fabs(Y2))+E-dE)
                     /(1e-3*e_0)));
 Y1=Y2;
}

}           /* end main */
```

Function definitions below this line

```
float
psi_at_inf(E)

float    E;                 /* the energy E */
{
 float   dz=1e-10;          /* the step length delta z */
 float   psi0,psi1,psi2;    /* psi(z-dz), psi(z), and psi(z+dz) */
 float   z;                 /* the spatial coordinate */
```

```
psi0=0;psi1=1;                          /* set starting values */
for(z=dz;z<100e-10;z+=dz)               /* first value of z is delta z */
{
  psi2=(2*m*(dz/hbar)*(dz/hbar)*
       (e_0*(z/100e-10)*(z/100e-10)-E)+2)*psi1-psi0;
  psi0=psi1;
  psi1=psi2;
}

return(psi2);
}
```

Note the include statement on the first line of the program, this is necessary in order to make use of the built-in function `fabs()` which returns the `floating point absolute value` (the modulus) of the argument.

As a quick test of its validity, the above program was used to solve the same parabolic potential as earlier, the results of which are given in Table 1.2. It can be seen that in the majority of cases there is a small improvement in the prediction of the energy separation $\hbar\omega = 39.039$ meV for the separation between adjacent eigenstates. In the next section it will be shown that the accuracy of these calculations is more dependent upon some of the choices of the computational parameters rather than the sophistication of the numerical methods themselves.

Proof
of Improved
Solutions

Table 1.2   Energy eigenvalues of the odd-parity solutions of a parabolic potential

n	step routine		mid-point rule	
	E_n (meV)	$E_n/(n-\frac{1}{2})$	E_n (meV)	$E_n/(n-\frac{1}{2})$
2	57.5	38.333	58.541	39.027
4	136.5	39.000	136.489	38.997
6	213.5	38.818	214.338	38.971
8	291.5	38.867	292.084	38.945
10	368.5	38.789	369.737	38.929
12	447.5	38.913	447.286	38.894

1.9 FURTHER CONVERGENCE TESTS: PUSHING ACCURACY TO ITS LIMITS

Three computational parameters have been used without real justification:

1. The step length δz (`dz` in the programs) for the iteration (the shoot forward) along the spatial axis

2. The energy step ΔE (dE in the programs) at which the value of the function $\psi(\infty, E)$ is sampled

3. The effective infinity, i.e. the distance along the spatial axis at which the iteration is halted (just defined numerically in the programs as the upper limit of the for loop over z, taken as 100 Å so far)

Consider the first of these—the effect on the energy eigenvalues of the step length δz. Table 1.3 shows the results of the same series of calculations as in Section 1.8, as calculated with the improved mid-point approximation, but this time as a function of the δz too.

Table 1.3 The effect on the energy eigenvalues of the value of the step length δz, the other two computational parameters are kept constant: $\Delta E = 1$ meV and the effective infinity equal to 100 Å

	$\delta z = 1$ Å		$\delta z = 0.5$Å		$\delta z = 0.2$ Å		*Further Improvements by Reducing Step Length*
n	E_n (meV)	$E_n/(n - \frac{1}{2})$	E_n (meV)	$E_n/(n - \frac{1}{2})$	E_n (meV)	$E_n/(n - \frac{1}{2})$	
2	58.541	39.027	58.564	39.043	58.571	39.047	
4	136.489	38.997	136.606	39.030	136.639	39.040	
6	214.338	38.971	214.625	39.023	214.704	39.037	
8	292.084	38.945	292.620	39.016	292.766	39.035	
10	369.737	38.929	370.590	39.009	370.825	39.034	
12	447.286	38.894	448.535	39.003	448.880	39.033	

When compared with the analytical value $\hbar\omega = 39.039$ meV for the energy spacing between the levels of the parabolic potential, it can be seen that the numerical predictions are in remarkable agreement. The lowest state in this series of calculations actually has the largest discrepancy, but even this equates to just 8 parts in 39,000, which is equivalent to 0.02%.

Table 1.4 The effect on the energy eigenvalue ($n = 2$ state) of the value of ΔE, the other two computational parameters are kept constant: $\delta z = 0.2$ Å and the effective infinity equal to 100 Å

ΔE (meV)	E_n (meV)	$E_n/(n - \frac{1}{2})$
1	58.571	39.047
0.5	58.558	39.039
0.2	58.557	39.038
0.1	58.557	39.038

Consider now the value of the energy step ΔE. Table 1.4 shows the result of decreasing the value of ΔE on the lowest energy odd-parity state. The second row of data in the table shows that sampling the value of the function $\psi(\infty, E)$ at 0.5 meV intervals allows the function $\psi(\infty, E) = 0$ to be solved to an accuracy of 0.001 meV, again a remarkable result.

Finally, consider the effect on the lowest odd-parity energy eigenvalue of the value of the effective infinity, this is summarised by the data in Table 1.5, but perhaps more vividly illustrated by Fig. 1.16. In any computational solution to a problem it is important that the results are not dependent upon the choice of the computational parameters. In this case it can be seen from the figure that any value of the 'effective infinity' above 60 Å will produce an acceptable result, this is confirmed in more detail by the data in Table 1.5.

Table 1.5   The effect on the energy eigenvalue ($n = 2$ state) of the value of the effective infinity, the other two computational parameters are kept constant: $\delta z = 0.2$ Å and $\Delta E = 0.1$ meV

'infinity' (Å)	E_n (meV)	$E_n/(n - \frac{1}{2})$
30	65.205	43.470
40	58.979	39.319
50	58.565	39.043
60	58.557	39.038
80	58.557	39.038
100	58.557	39.038
120	58.557	39.038
140	58.557	39.038

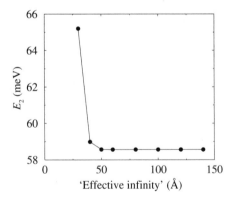

Fig. 1.16   The effect on an energy eigenvalue of the value of the 'effective infinity'

1.10 DEDUCTION OF THE WAVE FUNCTIONS

The algorithms and computer program solutions of Schrödinger's equation have so far only produced the energy eigenvalues. In many applications it is necessary to know the wave function as well. For example, some physical observables such as optical absorption and scattering cross-sections (probability that an electron collides with a phonon, for example) depend directly on the probability distribution (proportional to the square of the modulus of the wave function).

It is relatively easy to extend the technique to generate the wave function and in terms of the algorithmic description this just involves adding an extra statement after the energy eigenvalue has been determined accurately. In C this may appear as:

C Code Shooting Method with Extension to Generate Wave Functions

```
#include <math.h>

#define hbar 1.05459e-34
#define m 9.109534e-31
#define e_0 1.602189e-19
main()
{
float    psi_at_inf();      /* function psi(infinity,E) */
float    dE=0.01e-3*e_0;    /* the energy increment */
float    dummy;             /* dummy variable */
float    E;                 /* the energy E */
float    E_soln;            /* the solutions for the energy E */
float    Y1,Y2;             /* psi(infinity,E1), psi(infinity,E2) */

Y1=psi_at_inf(0.0,0);       /* evaluate first value */

for(E=dE;E<e_0;E+=dE)       /* start loop at small value of energy */
{
 Y2=psi_at_inf(E,0);        /* calculate next value */
 if(Y1*Y2<0)
 {
  E_soln=fabs(Y1)*dE/(fabs(Y1)+fabs(Y2))+E-dE;
  printf("E=%fmeV\n",E_soln/(1e-3*e_0));
  dummy=psi_at_inf(E,1);              /* print out wave function */
 }
 Y1=Y2;
}

}          /* end main */
```

Function definitions below this line

The Function $\psi(\infty, E)$ and the Code to Print $\psi(z)$

```
float
psi_at_inf(E,flag)
```

```
float    E;                      /* the energy E */
int      flag;                   /* wave function flag, if set print */
{
  float  dz=0.2e-10;             /* the step length delta z */
  float  psi0,psi1,psi2;         /* psi(z-dz), psi(z), and psi(z+dz) */
  float  z;                      /* the spatial coordinate */

  psi0=0;psi1=1;                             /* set starting values */
  if(flag)printf("0.000000e-10 %f\n",psi0);          /* first point */
  for(z=dz;z<60e-10;z+=dz)       /* first value of z is delta z */
  {
    if(flag)printf("%e %f\n",z,psi1);        /* print wf */
    psi2=(2*m*(dz/hbar)*(dz/hbar)*
         (e_0*(z/100e-10)*(z/100e-10)-E)+2)*psi1-psi0;
    psi0=psi1;
    psi1=psi2;
  }

return(psi2);
}
```

Notice that the function call to `psi-at-inf` contains an additional argument, which takes the value of 0 or 1. This second argument is captured by the variable `flag` in the function and, if equal to 1, causes the wave function to be printed. This just allows dual use of the function and minimises the number of lines of code for this simple illustration. In fact for a real computational code that was to be used frequently, the wave function output routine would not be written in this manner, because the additional `if` statements have to be evaluated many times as the energy solution is sought. This will slow down the operation of the code. For computational speed it is better to have more lines of code and separate the wave function generation to another function.

Note also the increased accuracy provided by the smaller energy step length $\Delta E = 0.01$ meV. Although this only allows the energy eigenvalue to be specified to a few more decimal places, which may often seem unnecessary, such an increased accuracy *is* necessary in order to produce a reasonable wave function. This can be understood by reflecting on Fig. 1.14 again—the value of $\psi(\infty, E)$ quickly takes on very large values at only small energies away from the solution.

Fig. 1.17 illustrates the wave functions (for $z > 0$ only) for the two lowest energy odd-parity solutions to the parabolic potential, i.e. the wave functions corresponding to the energy eigenvalues E_2 and E_4 of Table 1.3. Note that the wave functions tend *smoothly* to zero at the largest value of the spatial coordinate z (the effective infinity). The latter was chosen as a point at the beginning of the plateau of data in Fig. 1.16 and shows that the solutions do satisfy the 'standard boundary conditions' discussed earlier in Section 1.3.

In its present form the algorithm just generates *unnormalised* wave functions, which means that while they are eigenvectors of Schrödinger's equation, they are in fact some scalar multiple ($\zeta\psi$) of the *normalised* wave functions (ψ) which would satisfy the probability condition given in equation (1.69). The scalar multiple ζ does not alter the eigenvalue

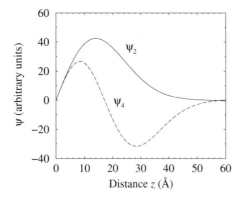

Fig. 1.17 The wave functions of the two lowest energy odd-parity solutions of the parabolic potential

in any way and should the wave functions be required as input to any further calculation then they can be easily normalised just by dividing by ζ, where:

$$\zeta^2 = \int_{-\infty}^{+\infty} \psi_u^*(z)\psi_u(z) \ dz \qquad (1.115)$$

the subscript 'u' just indicates that the integral over all space of the modulus is of the *u*n-normalised wave functions.

1.11 THREE-DIMENSIONAL POTENTIALS

Real physical systems are often characterised by general three-dimensional potentials. However, in this treatise, attention will be limited to systems with spherical symmetry, the reason being that this allows the Schrödinger equation to be deconstructed down into just one degree of freedom and hence a simple adaptation of the methods presented here can be used to solve it.

Consider again the Schrödinger equation for a particle in a three-dimensional potential:

$$-\frac{\hbar^2}{2m^*}\left[\frac{\partial^2}{\partial x^2} + \frac{\partial^2}{\partial y^2} + \frac{\partial^2}{\partial z^2}\right]\psi(x,y,z) + V(y,z)\psi(x,y,z) = E\psi(x,y,z) \quad (1.116)$$

Under the restraint of a spherically symmetric potential, i.e. $V(x,y,z) = V(r)$, where $r^2 = x^2 + y^2 + z^2$ then the wave function ψ will also have spherical symmetry, i.e. $\psi(x,y,z) = \psi(r)$, hence equation (1.116) becomes:

$$-\frac{\hbar^2}{2m^*}\left(\frac{\partial^2}{\partial x^2} + \frac{\partial^2}{\partial y^2} + \frac{\partial^2}{\partial z^2}\right)\psi(r) + V(r)\psi(r) = E\psi(r) \qquad (1.117)$$

The next step is to convert the Cartesian form of the ∇^2 into spherical polar coordinates. This is straightforward, though a little lengthy, and can be found in Appendix B. With this translation, equation (1.117) becomes:

Spherically Symmetric Schrödinger Equation

$$-\frac{\hbar^2}{2m^*}\left(\frac{2}{r}\frac{\partial}{\partial r}+\frac{\partial^2}{\partial r^2}\right)\psi(r)+V(r)\psi(r)=E\psi(r) \tag{1.118}$$

Such spherically symmetric Schrödinger equations have been investigated before by well established techniques[7]. As an alternative and with the impetus in this work on simple numerical schemes, a shooting technique similar to that described for one-dimensional potentials is sought. With this aim, expanding the first and second derivatives in terms of finite differences gives:

$$\frac{2}{r}\left[\frac{\psi(r+\delta r)-\psi(r-\delta r)}{2\delta r}\right]+\frac{\psi(r+\delta r)-2\psi(r)+\psi(r-\delta r)}{(\delta r)^2}$$

$$=\frac{2m^*}{\hbar^2}\left[V(r)-E\right]\psi(r) \tag{1.119}$$

Multiplying both sides by $r(\delta r)^2$ gives:

$$\left[\psi(r+\delta r)-\psi(r-\delta r)\right]\delta r+r\left[\psi(r+\delta r)-2\psi(r)+\psi(r-\delta r)\right]$$

$$=r(\delta r)^2\frac{2m^*}{\hbar^2}\left[V(r)-E\right]\psi(r) \tag{1.120}$$

Gathering terms in $\psi(r+\delta r)$, $\psi(r)$, and $\psi(r-\delta r)$ then:

$$(r+\delta r)\,\psi(r+\delta r)$$

$$=r\left\{(\delta r)^2\frac{2m^*}{\hbar^2}\left[V(r)-E\right]+2\right\}\psi(r)+(-r+\delta r)\,\psi(r-\delta r) \tag{1.121}$$

Spherically Symmetric Shooting Equation

and finally:

$$\psi(r+\delta r)=\frac{r\left\{2m^*(\delta r/\hbar)^2\left[V(r)-E\right]+2\right\}\psi(r)+(-r+\delta r)\,\psi(r-\delta r)}{r+\delta r} \tag{1.122}$$

which is again an iterative equation which can be solved with a numerical shooting technique according to the standard boundary conditions as discussed in Section 1.3.

1.12   APPLICATION TO THE HYDROGEN ATOM

Consider the electron-proton system that constitutes a hydrogen atom. The relatively light electron moves in the potential of the much heavier proton and, as a first approximation, the position of the proton is often considered fixed. The potential energy of the electron due to the Coulombic interaction that the electron experiences is spherically symmetric and given by:

$$V(r)=-\frac{e^2}{4\pi\epsilon_0 r} \tag{1.123}$$

Now this system can be solved analytically either with the semi-classical approach of Bohr[8] or fully quantum mechanically. However, it suits the purpose here to consider

---

[7]See for example Schiff, *Quantum Mechanics* (McGraw-Hill, London 1968).
[8]See any modern physics book.

numerical solutions as an example and a way of validating the approach derived above for spherically symmetric potentials. A C implementation of the solution might appear as:

C Code for Numerical Solution of Hydrogen

```
#include <math.h>

#define hbar 1.05459e-34
#define m 9.109534e-31
#define e_0 1.602189e-19
#define epsilon_0 8.854188e-12
#define pi 3.141593
main()
{
float    psi_at_inf();     /* function psi(infinity,E) */
float    dE=1e-3*e_0;      /* the energy increment */
float    dummy;            /* dummy variable */
float    E;                /* the energy E */
float    E_soln;           /* the solutions for the energy E */
float    Y1,Y2;            /* psi(infinity,E1), psi(infinity,E2) */

Y1=psi_at_inf(-20*e_0,0);          /* evaluate first value */


for(E=-20*e_0+dE;E<0;E+=dE)        /* start loop at small value of energy */
{
 Y2=psi_at_inf(E,0);       /* calculate next value */
 if(Y1*Y2<0)
 {
  E_soln=fabs(Y1)*dE/(fabs(Y1)+fabs(Y2))+E-dE;
  printf("E=%fmeV\n",E_soln/(1e-3*e_0));
  dummy=psi_at_inf(E,1);              /* print out wave function */
 }
 Y1=Y2;
}

}          /* end main */
```

Function definitions below this line

```
float
psi_at_inf(E,flag)
```

$\psi(\infty, E)$

```
float    E;                /* the energy E */
int      flag;             /* wave function flag, if set print */
{
 float    V();              /* the potential energy function */
 float    dr=0.001e-10;     /* the step length delta z */
 float    psi0,psi1,psi2;   /* psi(z-dz), psi(z), and psi(z+dz) */
 float    r;                /* the spatial coordinate */
```

```
psi0=1;psi1=1;                          /* set starting values */
if(flag)printf("0.000000e-10 %f\n",psi0);        /* first point */
for(r=dr;r<10e-10;r+=dr)            /* first value of z is delta z */
{
  if(flag)printf("%e %f\n",r,psi1);        /* print wf */
  psi2=(r*(2*m*(dr/hbar)*(dr/hbar)*(V(r)-E)+2)*psi1+
       (-r+dr)*psi0)/
       (r+dr);
  psi0=psi1;
  psi1=psi2;
}

return(psi2);
}
```

Function Defining Coulombic Potential Energy

```
float
V(r)

float    r;                     /* the spatial coordinate */
{
  return(-e_0*e_0/(4*pi*epsilon_0*r));
}
```

Note that in this evolution the potential, now called $V(r)$, has been separated into a function—this just allows for easier reading and subsequent editing.

There are several points to note about the computational implementation above. In particular, the form of the potential gave some clues as to sensible choices for parameters. For example, the potential clearly diverges as $r \to 0$ so it was decided to shoot the wave function forward from a point (dz) very close to the origin. Relevant experience was utilised: it is known that the energy eigenvalues of the hydrogen atom are separated by electron volts, so a choice for the energy step dE of 1 meV would always be sufficient to separate the solutions. The first (lowest) energy to sample $\psi(\infty, E)$ was initially chosen at a very large and negative value: some calculations showed that the eigenvalues were all above -20 eV. As usual, the value for the effective infinity was chosen large enough so as not to affect the lowest energy eigenvalue. Perhaps the most important feature of the method, which is implemented in the code and has not yet been mentioned, are the starting values for the iteration over the spatial coordinate, i.e. $\psi(0)$ and $\psi(\delta r)$. Given the information about the problem presented here, then these values are not known and use cannot be made of the knowledge of odd- and even-parity states. In fact the starting values were 'guessed' and tried as '1' and '1', and they worked (see below)—this, together with a similar example of unexpected starting conditions[9] goes some way to demonstrating the stability and ruggedness of the method.

The results of the calculations are shown in Table 1.6.

[9] See Harrison, *Quantum Wells, Wires and Dots*, (Wiley, Chichester, 1999) p.75.

Table 1.6   The energy eigenvalues of a hydrogen atom calculated with $\delta z = 0.001$ Å, $\Delta E = 1$ meV and the effective infinity was chosen as 10 Å

Comparison Between Numerical and Accepted Results

n	E_n (eV)	Bohr theory (eV)
1	-13.607	-13.606
2	-3.405	-3.402
3	-1.289	-1.512

The first column indicates the principal quantum number (or label) of the energy eigenvalues. There is no odd- and even-parity as in the case of the one-dimensional parabolic potential, so the lowest state calculated really is the ground state of the system. The second column gives the energy eigenvalues as calculated here and these are compared with values predicted from the simple Bohr theory of the hydrogen atom[10].

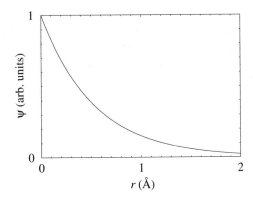

Fig. 1.18   The ground state wave function of the hydrogen atom

It can be seen that the ground state energy eigenvalue is in excellent agreement with Bohr theory, as is the first excited ($n = 2$) level, which follows as $E_1/2^2$. However, the energy of the second excited state ($n = 3$) is not in good agreement with Bohr theory ($E_1/3^2$), and this is probably because the value of the effective infinity is too small. It is left as an exercise for the reader to verify this.

Fig. 1.18 shows the radial component of the wave function, i.e. $\psi(r)$ as a function of the distance r of the electron from the fixed proton nucleus. It can be seen that, although the starting conditions for the iteration along this axis where taken as '1' and '1', the wave function seems to just decay exponentially. This can occur because the mesh—the points

---

[10]Again, see any textbook on modern physics or quantum theory.

at which the wave function is calculated, are quite close together ($\delta z = 0.001$ Å in this calculation) and hence the initial 'flat' starting conditions are hidden as detail.

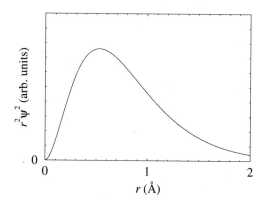

Fig. 1.19   The probability density of the ground state of the hydrogen atom

Fig. 1.19 illustrates the physics of this example hydrogen atom calculation somewhat more clearly. It plots the probability of finding the electron in the spherical shell of space between the distances r and $r + \delta r$ from the proton. The figure shows that the electron has the highest probability of being 0.529 Å from the proton. Bohr theory calls this distance the 'Bohr radius' and calculates it to be 0.528 Å. Again, the numerical calculations here are in excellent agreement.

Summary

The total energy of a quantum system was shown to be represented by the eigenvalues of its Hamiltonian operator, which in one dimension is given by:

$$\mathcal{H} = -\frac{\hbar^2}{2m}\frac{\partial^2}{\partial z^2} + V(z)$$

where different quantum systems are specified by the potential V. Using a finite difference approximation for the second derivative, the corresponding Schrödinger equation was rearranged into an iterative shooting equation:

$$\psi(z + \delta z) = \left\{ \frac{2m}{\hbar^2}(\delta z)^2 [V(z) - E] + 2 \right\} \psi(z) - \psi(z - \delta z)$$

The starting values were shown to be:

$$\psi(0) = 0 \quad \text{and} \quad \psi(\delta z) = 1 \quad \text{for odd-parity states}$$

and

$$\psi(0) = 1 \quad \text{and} \quad \psi(\delta z) = \frac{m}{\hbar^2}(\delta z)^2 [V(0) - E] + 1 \quad \text{for even-parity states}$$

The energy eigenvalues E are found by searching for wave functions $\psi(z)$ which tend smoothly to zero as the spatial coordinate z tends to infinity.

Tasks

1.1 **Infinitely deep quantum wells:** By imposing the standard boundary conditions on a constant (zero) potential surrounded by infinitely high barriers, deduce the analytical forms for the energy eigenvalues and the *normalised* wave functions. Write a computer program to evaluate the first few eigenvalues and generate their associated wave functions. Plot the first four wave functions.

1.2 **Finite quantum well:** For the finite-quantum-well potential given by:

$$V(z) \;=\; V, \qquad z < -\frac{a}{2} \tag{1.124}$$

$$V(z) \;=\; 0, \qquad -\frac{a}{2} \leq z \leq +\frac{a}{2} \tag{1.125}$$

$$V(z) \;=\; V, \qquad +\frac{a}{2} < z \tag{1.126}$$

show that the wave function in the well region ($V = 0$) can be written as:

$$\psi(z) = A \cos kz + B \sin kz \tag{1.127}$$

and in the right-hand barrier as:

$$\psi(z) = C \exp{(-\kappa z)} \tag{1.128}$$

where k and κ are given in Section 1.3. By equating the wave function and its derivative at the interface between the well and the right-hand barrier, show that the energy eigenvalues of the even-parity states are solutions to the equation:

$$k \tan\left(\frac{ka}{2}\right) = \kappa \tag{1.129}$$

This is an equation with a single unknown—the energy E. Write a computer program to find the solutions for E.

1.3 **Triangular quantum wells:** Find the analytical solutions to the infinitely deep triangular well described in Section 1.3. Write a computer program, in a language of your choice, to generate the first few eigenvalues. By looking up the functional forms for the Airy functions Ai and Bi (see, for example, Abramowitz and Stegun, *Handbook of Mathematical Functions*, (Dover, New York, 1965)), write a program to calculate (and hence plot) the corresponding wave functions.

1.4 **Pöschl-Teller potential holes:** By plotting the Pöschl-Teller potential for a range of λ (and a fixed value of α), deduce the domain for λ which gives bound solutions, i.e. $E < 0$.

1.5 **Shooting method for the even parity states of a parabolic potential:** Adapt the C-sourcecode for the odd-parity solutions of the parabolic potential, as given in Section 1.7, to generate the *even-parity* eigenvalues.

1.6 **Shooting method for finite-quantum-wells:** Implement the shooting method for the eigenvalue solutions of the finite-quantum-well potential of Task 1.2, in a programming language or environment of your choice. What happens to the energy eigenvalues as the width of the quantum well increases? What happens when the height of the barrier increases?

1.7 **The value of the 'effective infinity':** The effective infinity, i.e. the distance which the solutions are shot forward along the z-axis was chosen arbitrarily, but shown to give answers which agree well with analytical solutions. Taking the parabolic quantum well, and using the C sourcecode in Section 1.8 or otherwise, deduce the range of values for the effective infinity which gives energy eigenvalues that agree well with the analytical form in equation (1.75). Is this range different for the lower energy states than the higher? How can it be ensured that a good choice is made for the effective infinity?

1.8 **Convergence tests for the parabolic potential:** Using the C sourcecode in Section 1.8, increase the numerical accuracy by way of both the spatial and energy step sizes, δz and δE respectively. See how accurately the analytical value for $\hbar\omega$ can be reproduced.

1.9 **Spherical quantum dots:** Interesting new semiconductor devices are being made from nanometre size dots of material. One approximation to the dots is to describe them as spheres with a constant potential within the dot and a constant (but higher) potential outside of the dot. Adapt the shooting method for spherically symmetric potentials to describe these systems and hence deduce the energy eigenvalues of electrons in semiconductor quantum dots made from GaAs surrounded by $Ga_{0.8}Al_{0.2}As$. *(Assume the electron mass is constant and equal to $0.067m_0$ and the conduction band offset is 167 meV.)*

Projects

1.1 **Shooting method for a general potential:** Write a computer program that reads in a general symmetric potential given as a two-column list of z and $V(z)$ in a file and then solves the corresponding one-dimensional Schrödinger equation, for both even and odd parity bound ($E < V_{max}$) states.

1.2 **Non-symmetric potentials:** A substantial fraction of potentials of interest are not symmetric, for example, the one-dimensional quantum well potential under the influence of a linear electric field. The consequence of this is that the eigenvector solutions (the wave functions) do not have exact even- or odd-parity, hence the starting conditions derived in Section 1.5 are inappropriate. By considering the physics of the expected wave function solutions, deduce starting conditions for the shooting method. Prove their worth by comparing the energy eigenvalues and wave functions from this more generalised shooting technique with the asymmetric triangular quantum well potential discussed in Task 1.3.

1.3 **Numerical solution of finite-quantum-wells:** Extend the finite-quantum-well eigenvalue solution implemented in Task 1.6 to generate the wave functions of both the even- and odd-parity states. In semiconductor systems the movement of electrons can be characterised in terms of an effective mass m^*. Finite-quantum-well potentials can be formed by alternating thin layers of dissimilar semiconductors, which by definition have different electron effective masses; given that the solutions should now conform to:

$$\psi \quad \text{and} \quad \frac{1}{m^*}\frac{d\psi}{dz} \quad \text{are continuous}$$

show that the energy eigenvalues of the even-parity states are given by:

$$\frac{k}{m_w^*}\tan\left(\frac{ka}{2}\right) = \frac{\kappa}{m_b^*} \qquad (1.130)$$

where m_w^* and m_b^* are the electron effective masses in the well and barrier region respectively. Deduce the corresponding equation for the odd-parity states. The resulting equation contains a single unknown—the energy eigenvalue E. Implement a numerical solution for both the energies and wave functions of the confined states.

Compare the results with those produced by the shooting method (the latter requires a reworking of the shooting equation due to the altered form of the Hamiltonian that the new boundary conditions represent, see Harrison, 'Quantum Wells, Wires and Dots, (Wiley, Chichester, 1999) for the new equation).

1.4 **Improved energy eigenvalue solutions from the shooting method:** Section 1.8 shows how to improve the estimate of the energy eigenvalue solutions of Schrödinger's equation by supplementing the simple step-routine with a mid-point rule. Develop a Newton-Raphson iteration to improve the solution of $\psi(\infty, E) = 0$ still further and demonstrate its accuracy by incorporating it within the C-sourcecode of Section 1.8 (or a language or environment of your choice) and applying it to systems with analytical solutions, such as the parabolic potentials, Pöschl-Teller potential hole and finite-quantum-wells.

1.5 **Higher-order potentials:** The solutions to parabolic (z^2) potentials have been extensively studied. Using the shooting method, solve potentials of the form z^{2n}, where n is a positive integer. Plot these potentials and hence deduce the analytical solution to the potential:

$$V = \lim_{n \to \infty} \frac{C}{2} z^n \tag{1.131}$$

where C is some constant. Compare the result with the numerical solutions in the limit of large n.

Potentials of the form $az^3 + bz^2 + c$ do not have bound states in the true sense, however, the *local* minima do allow for 'quasi-bound states'. By choosing suitable values for a, b and c, generate a potential with a local minimum and show how the solutions for the quasi-bound states can be found. Adapt a suitable computer program to illustrate this.

1.6 **High-energy physics:** Yukawa proposed the existence of a charged particle, called the meson, which carried the nuclear force binding neutrons to protons. The meson moves within the nucleus under the influence of a screened Coulomb potential, now often known as the 'Yukawa potential':

$$V(r) = -\frac{Ze^2}{4\pi\epsilon_0 r} e^{-r/a}$$

Given that the mass of a meson (the π-meson or pion) is about 270 times the mass of the electron and that the effective extent of the potential (determined by the value of the decay constant a) is of the order of the size of a nucleus, calculate the energy levels of mesons within a nucleus consisting of 10 protons and 10 neutrons.

1.7 **Quark–anti-quark interactions:** Killingbeck *Microcomputer Quantum Mechanics*, (Adam Hilger, Bristol, 1983) p.150 gives the Schrödinger equation for the bound quark–anti-quark interaction in the form of:

$$-\frac{\hbar^2}{2\mu}\nabla^2\psi - \frac{\beta}{r}\psi + \gamma r^M \psi = E\psi$$

where μ is the reduced mass of the system, β and γ are parameters and M is a positive integer. *(Note that the ever increasing potential described by the power function r^M implies that the quark–anti-quark pair will always be bound.)* Find typical values for quark masses and, with β set equal to 1, find values of the parameter γ that give energies of the typical order, probed with modern particle accelerators, for different values of M.

2

Approximate Methods

2.1 SMALL CHANGES

Often the solutions of a physical system are known, or have been deduced, from some intensive numerical calculation, and the effect of some small change (or perturbation) to the system would like to be evaluated. For example, the clock, usually referred to as 'Big Ben' in the Palace of Westminster in London is controlled by a pendulum, the periodic time of which is adjusted by adding and removing very small masses, actually Victorian pennies—*it must be a compound pendulum*. The new periodic time is clearly quite close to the original period, so it would be convenient if the new time could be deduced from the original time rather than resorting to a full new solution. The pinnacle of motorsport—Formula 1 is becoming increasingly analytical, and complex calculations are performed to predict the laptime of the cars under various racing conditions such as air temperature (which effects the aerodynamics) and track condition (a wet track reduces grip on the bends). Given the great complexity of the calculations, experimental data is fed into the computer models and the parameter space can be explored by considering changes to the car and track as perturbations, for example, the laptime when the car has only a half fuel load might be derived from the laptime *measured* when the fuel tank was full. This would be achieved by calculating the effect of the change in mass of the car on the known (experimentally measured) solution, rather than resorting to a full (and very difficult) first-principles solution.

In this instance, attention is going to be focussed on quantum mechanical systems and there are many examples where solutions to well known effects can be found by considering them as a perturbation, i.e. a small change in the potential. For example, in the Zeeman effect a magnetic field is applied to a system of atoms and the electron energy levels are split into a series of levels arising from the alignment of the angular momentum along the direction of the field. Another example is evaluation of the effect of an electric field on the energy levels of electrons within semiconductor quantum wells, as found in many modern optoelectronic devices.

'Non-degenerate' perturbation theory is an approach for evaluating the effect of a small additional potential on a system where as the name suggests none of the states have the same energy. The counterpart does exist: 'Degenerate' perturbation theory tackles the small changes in the potential for systems with states having the same energy (see Section 3.8).

Let an original quantum system be described by some Hamiltonian $\mathcal{H}_0$ and say all solutions, i.e. the energy eigenvalues e_i and eigenvectors (wave functions) ψ_i, are known:

$$\mathcal{H}_0 \psi_i = e_i \psi_i \tag{2.1}$$

Then the eigenfunctions ψ_i form a complete orthonormal set, which means that any other function can be expressed as a linear combination of them. This is rather like the construction of any periodic function as a linear combination of sine waves in Fourier analysis.

If the system is perturbed by some *small* additional potential, then the new Hamiltonian describing it is equal to the original Hamiltonian plus some additional terms, i.e.

$$\mathcal{H} = \mathcal{H}_0 + \mathcal{H}' \tag{2.2}$$

and the perturbed quantum system has new solutions:

$$\mathcal{H}\Psi_I = E_I \Psi_I \tag{2.3}$$

If the new Hamiltonian cannot be solved exactly, either analytically or numerically (as discussed in Chapter 1) then one alternative is to find approximate solutions for the new energy eigenvalues E_I and wave functions Ψ_I. The method which follows below is one such approach, known as 'perturbation theory'.

2.2 NON-DEGENERATE PERTURBATION THEORY

Using the form for the perturbed Hamiltonian in equation (2.2) in the time-independent Schrödinger equation given in equation (2.3):

Perturbed Schrödinger's Equation

$$(\mathcal{H}_0 + \mathcal{H}') \Psi_I = E_I \Psi_I \tag{2.4}$$

Now the solution Ψ_I to the perturbed system can clearly be expressed as a linear combination of the unperturbed solutions ψ_j:

Expand Wave Function

$$\Psi_I = \sum_{j=1}^{n} a_j \psi_j \tag{2.5}$$

Hence, substituting into equation (2.4), gives:

$$\sum_{j=1}^{n} a_j \mathcal{H}_0 \psi_j + \mathcal{H}' \Psi_I = E_I \sum_{j=1}^{n} a_j \psi_j \tag{2.6}$$

Multiplying by ψ_i^* and integrating over all space:

$$\sum_{j=1}^{n} a_j \int \psi_i^* \mathcal{H}_0 \psi_j \ d\tau + \int \psi_i^* \mathcal{H}' \Psi_I \ d\tau = E_I \sum_{j=1}^{n} a_j \int \psi_i^* \psi_j \ d\tau \qquad (2.7)$$

Now ψ_j is an eigenfunction of $\mathcal{H}_0$, with an eigenvalue e_j (see equation (2.1)), hence:

$$\sum_{j=1}^{n} a_j \int \psi_i^* e_j \psi_j \ d\tau + \int \psi_i^* \mathcal{H}' \Psi_I \ d\tau = E_I \sum_{j=1}^{n} a_j \int \psi_i^* \psi_j \ d\tau \qquad (2.8)$$

Recalling that the solutions ψ_j form a complete orthonormal set, i.e. the integral over all space of $\psi_i^* \psi_j$ is equal to the Kronecker-delta δ_{ij}, then:

$$\sum_{j=1}^{n} a_j e_j \delta_{ij} + \int \psi_i^* \mathcal{H}' \Psi_I \ d\tau = E_I \sum_{j=1}^{n} a_j \delta_{ij} \qquad (2.9) \qquad \text{\textit{Use Orthonormality}}$$

The Kronecker deltas limit the contribution of the summations to the single term when $j = i$, therefore:

$$a_i e_i + \int \psi_i^* \mathcal{H}' \Psi_I \ d\tau = E_I a_i \qquad (2.10)$$

Now the energy eigenvalue of the 'I'th level of the perturbed system, E_I, must be equal to the energy of the equivalent level of the unperturbed system e_i plus some change, i.e. $E_I = e_i + \Delta e_i$, hence:

$$a_i e_i + \int \psi_i^* \mathcal{H}' \Psi_I \ d\tau = (e_i + \Delta e_i) a_i \qquad (2.11)$$

Then:

$$a_i \Delta e_i = \int \psi_i^* \mathcal{H}' \Psi_I \ d\tau \qquad (2.12)$$

But the problem remains as to what the perturbed wave function Ψ_I is. This new wave function was taken as a linear combination of the unperturbed eigenstates as in equation (2.5) which amounts to:

$$\Psi_I = a_0 \psi_0 + a_1 \psi_1 + \ldots + a_{i-1} \psi_{i-1} + a_i \psi_i + a_{i+1} \psi_{i+1} + \ldots + a_n \psi_n \qquad (2.13)$$

Consider approximating Ψ_I by the single term $a_i \psi_i$, which amounts to the contribution from the original unperturbed state, then equation (2.12) becomes:

$$a_i \Delta e_i = a_i \int \psi_i^* \mathcal{H}' \psi_i \ d\tau \qquad (2.14)$$

The coefficient a_i can be divided through giving a first-order correction (label it with the superscript (1)) to the change in the energy eigenvalue of a state [1] :

First-order perturbative correction

$$\Delta e_i^{(1)} = \int \psi_i^* \mathcal{H}' \psi_i \ d\tau \tag{2.15}$$

Similar arguments[2] give the second-order correction as a sum of the matrix elements squared over all other states:

Second-order perturbative correction

$$\therefore \Delta e_i^{(2)} = -\sum_{j \neq i} \frac{|\int \psi_j^* \mathcal{H}' \psi_i \ d\tau|^2}{e_j - e_i} \tag{2.16}$$

The final approximation to the energy E_I of the perturbed state is found by simply adding both first-order and second-order contributions to the energy e_i of the original unperturbed state, i.e.

$$E_I \approx e_i + \Delta e_i^{(1)} + \Delta e_i^{(2)} \tag{2.17}$$

2.3 COMPUTING THE FIRST-ORDER CORRECTION

The infinitely-deep quantum well (discussed in Task 1 of Chapter 1) is a very convenient system in which to illustrate how to evaluate the effect of perturbations, because the energies and wave functions of the unperturbed states are simple analytical expressions. Thus concentration can be focussed upon the points of interest in this chapter. In particular, for a well of width L, as illustrated in Fig. 2.1, the energy eigenvalues are given by:

Infinitely Deep Quantum Well Energies

$$e_n = \frac{\hbar^2 \pi^2 n^2}{2mL^2} \tag{2.18}$$

and, with the origin set at the left-hand edge of the well, the wave functions are:

Infinitely Deep Quantum Well Wave Functions

$$\psi_n = \sqrt{\frac{2}{L}} \sin\left(\frac{\pi n z}{L}\right) \tag{2.19}$$

Evaluation of the first-order change in the energy of one of these one-dimensional infinitely-deep quantum well states, due to the effect of some perturbing potential, amounts only to the calculation of a simple one-dimensional integral, as given by equation (2.15). In this manifestation, the integral is going to be calculated by a simple strip summation, as illustrated in Fig. 2.2.

---

[1] Integrals of this type are often represented in a more compact form called 'Dirac notation'. The integral in equation (2.15) would be written as: $\langle \psi_i | \mathcal{H}' | \psi_i \rangle$ or even more succinctly just as $\langle i | \mathcal{H}' | i \rangle$. Such entities often occur as the elements of matrices and for that reason they are commonly known as 'matrix elements'. The integral in equation (2.16) is a matrix element between two different states and would follow as $\langle \psi_j | \mathcal{H}' | \psi_i \rangle$ or $\langle j | \mathcal{H}' | i \rangle$. The integral making up the Kronecker delta contains no operator and would be written as $\langle \psi_i | \psi_j \rangle$ or $\langle i | j \rangle$.
[2] See any good quantum mechanics book.

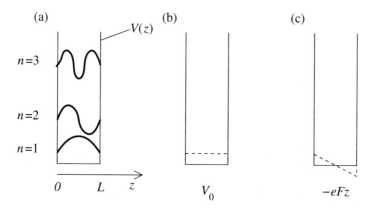

Fig. 2.1　(a) The infinite quantum well together with schematic representation of the three lowest energy solutions; (b) a constant perturbing potential and (c) a perturbing potential due to an electric field

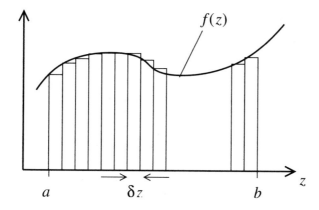

Fig. 2.2　A simple algorithm for evaluating definite integrals numerically

Mathematically this would be described as:

$$\int_a^b f(z)\ dz = \sum_{i=0}^{(b-a)/\delta z} f(a + i\delta z)\ \delta z \tag{2.20}$$

Note that in this form the area of each strip is taken as the height at the left-edge, multiplied by the width of the strip. Equivalent expressions could be constructed with the height evaluated at the centre of the strip or at the right-hand edge, but the latter appear to offer no advantage. It can be seen from Fig. 2.2 that in some regions the strip summation will over-estimate, and in others it will under-estimate, the area under the curve. More accurate means of evaluating definite integrals are discussed in Tasks 2.1 and 2.2, together with Project 2.1 at the end of this chapter.

An algorithmic description for evaluating the first-order correction utilising the simple strip method is given in Algorithm 2.1.

Algorithm 2.1 Evaluating the first-order perturbative correction

---

$L \Leftarrow 100 \times 10^{-10}$ {Define parameters}

$\delta z \Leftarrow 1 \times 10^{-10}$

$i \Leftarrow 1$

Cumulative-sum$\Leftarrow 0$ {Initialise variable for summation}

for z=some-lower-limit to some-upper-limit **do**

   Add $\psi_i \times \mathcal{H}'(z) \times \psi_i \times \delta z$ to Cumulative-sum

end for

Print final value of Cumulative-sum

---

where $\mathcal{H}'(z)$ is a function specifying the perturbing potential.

   Consider first the constant perturbation across the whole width of the quantum well, as illustrated in Fig. 2.1(b). The C code to compute the first-order correction may appear as:

C Code for Evaluating First Order Correction to Energy

```c
#include <math.h>

#define e_0 1.602189e-19
#define pi 3.141592

main()
{
double  Hprime();          /* the perturbation function */
double  dz;                /* the integration step width */
double  L=100e-10;         /* quantum well width */
double  sum=0;             /* the cumulative sum */
double  psi;               /* the wave function */
double  z;                 /* the spatial coordinate */
int     n=1;               /* the state of interest */

dz=L/100;
for(z=0;z<L;z+=dz)                     /* loop along spatial axis */
{
 psi=sqrt(2/L)*sin(pi*n*z/L);     /* calculate wave function */
 sum+=psi*Hprime(z)*psi*dz;       /* calculate area of strip */
}
printf("Delta_e_%i=%f eV\n",n,sum/e_0); /* output result */
}

double
Hprime(z)                      /* the perturbation */

double  z;
{
 return(1*e_0);
}
```

Table 2.1 shows the results of a series of calculations of the first-order change in energy $\Delta e_i^{(1)}$, for a range of constant perturbing potentials, for the ground ($i = 1$) and first-excited ($i = 2$) states of the infinitely deep quantum well.

Table 2.1 The first-order change in energy of the two lowest energy states in a ($L =$) 100 Å wide, infinitely-deep quantum well as a function of the magnitude V_0 of a constant perturbing potential

V_0 (eV)	Δe_1 (eV)	Δe_2 (eV)
1	1.000000	1.000000
2	2.000000	2.000000
5	5.000001	5.000001
10	10.000002	10.000002

The table shows that the first-order change in energy is equal to the magnitude of the perturbing potential. This is not surprising if due consideration is given to the mathematical form of the first-order correction. With the perturbation given by some constant potential which acts across the full range of integration, then the first-order correction in equation (2.23) becomes:

$$\Delta e_i^{(1)} = \int_0^L \psi_i^* V_0 \psi_i \; dz = V_0 \int_0^L \psi_i^* \psi_i \; dz \qquad (2.21)$$

given that the entire system is contained within the spatial limits $0 < z < L$, then the integral expression is just the wave function normalisation condition which, from equation (1.69), equals 1. Hence, for this type of perturbation, $\Delta e_i^{(1)} = V_0$, as found numerically. In this series of calculations 100 strips were used to calculate the definite integral. The data in the table shows that this produces very close agreement with the analytical solutions—this serves as evidence for the validity of the rather simple strip summation for evaluation of the definite integral.

An electric field produces a 'potential difference' between two points. Assuming the material is linear, isotropic and homogeneous, then the potential will be a linear function of position, i.e. in the one-dimensional examples met in Chapter 1:

$$V(z) = -eFz \qquad (2.22)$$

as illustrated in Fig. 2.1(c). Table 2.2 shows the results of calculations of the first-order corrections on the ground ($i = 1$) and first excited ($i = 2$) states of the same 100 Å infinitely-deep quantum well as in the previous example.

Again the results hint at some underlying analytical solution, and indeed this is the case. Consider the form for the first-order correction, with this linear perturbation:

$$\Delta e_i^{(1)} = \int \psi_i^* \left(-eFz \right) \psi_i \; dz \qquad (2.23)$$

Table 2.2 The first-order change in energy of the two lowest energy states in a ($L =$) 100 Å wide infinitely-deep quantum well as a function of the field F of a linear perturbing potential $(-eFz)$. Note that the origin for the field has been adjusted to lie at the centre of the well.

F (kVcm^{-1})	Δe_1 (eV)	Δe_2 (eV)
10	0.000000	0.000000
20	0.000000	0.000000
50	0.000000	0.000000
100	0.000000	0.000000

Chapter 1 has already demonstrated that the wave functions ψ_i of bound-states are real and, if the potential is symmetric, they are either symmetric or anti-symmetric, as illustrated for the single-quantum-well example in Fig. 2.1(a). The square of any of these wave functions (as occurs in equation (2.23)) is an even (symmetric) function. When combined with the odd (asymmetric) perturbing potential function $(-eFz)$, the first-order change in the energy $\Delta e_i^{(1)}$ is a definite integral of an asymmetric function—which when evaluated between equal and opposite limits is always zero. Thus, *the first-order change in the energy of a bound-state of a symmetric potential when under the influence of an electric field, is zero.* However, the second-order change is non-zero and this will be evaluated in the next section.

2.4 COMPUTING THE SECOND-ORDER CORRECTION

This change in the energy of a quantum state due to the application of an electric field is often referred to as the 'Stark effect', and the particular example of the suppression of the energy of a state in a quantum well is labelled the 'quantum-confined Stark effect'. In the previous section it has been shown that to first order, the change in energy of a state in a symmetric quantum well is zero, now, however, consider the second-order correction given in equation (2.16). The main function is to evaluate a series of integrals $\langle \psi_j | \mathcal{H}' | \psi_i \rangle$ in a manner similar to that in the first-order correction, and then to sum these contributions for a variety of states ψ_j—the algorithm (given below as Algorithm 2.2) can be considered as an extension to the previous with an additional loop over the basis states 'j'.

The C-code manifestation of this algorithm might look like:

C Code for Evaluating Second Order Correction to Energy

```
#include <math.h>

#define e_0 1.602189e-19
#define hbar 1.05459e-34
#define m 9.109534e-31
#define pi 3.141592
```

Algorithm 2.2 Evaluating the second-order perturbative correction

---

$L \Leftarrow 100 \times 10^{-10}$ {Define parameters}

$\delta z \Leftarrow 1 \times 10^{-10}$

$i \Leftarrow 1$

Cumulative-sum-over-states $\Leftarrow 0$ {Initialise variable for summation over states}

for j=2 to some-upper-state **do**

    Cumulative-sum $\Leftarrow 0$ {Initialise variable for summation}

    for z=some-lower-limit to some-upper-limit **do**

        Add $\psi_j \times \mathcal{H}'(z) \times \psi_i \times \delta z$ to Cumulative-sum

    end for

    Add $(\text{Cumulative-sum})^2 / (E_j - E_i)$ to Cumulative-sum-over-states

end for

Print final value of Cumulative-sum-over-states

```
main()
{
double   E();                    /* energy of the jth state */
double   Hprime();               /* the perturbation function */
double   dz;                     /* the integration step width */
double   L=100e-10;              /* quantum well width */
double   sum;                    /* the cumulative sum */
double   sum_state=0;            /* cumulative sum over states */
double   psi_i;                  /* the wave function */
double   psi_j;                  /* the wave function */
double   z;                      /* the spatial coordinate */
int      i=1;                    /* the state of interest */
int      j;                      /* index over states */
int      upper_state=10;   /* highest state in summation */


dz=L/100;
for(j=2;j<=upper_state;j++)
{
 sum=0;                          /* sum initialised for each j */
 for(z=0;z<L;z+=dz)              /* loop along spatial axis */
 {
  psi_i=sqrt(2/L)*sin(pi*i*z/L);              /* psi-i */
  psi_j=sqrt(2/L)*sin(pi*j*z/L);              /* psi-j */
  sum+=psi_j*Hprime(z)*psi_i*dz;              /* area of strip */
 }
 sum_state-=sum*sum/(E(L,j)-E(L,i));
 printf("Incl. %i states Delta_e=%f eV\n",j,sum_state/e_0);
}
}
```

Integral

$\Delta e^{(2)}$

```
double   E(L,j)                  /* the energy of the jth level */
double   L;
int      j;
{
 return(hbar*hbar*pi*pi*j*j/(2*m*L*L));
}

double   Hprime(z)               /* the perturbation */
double   z;
{
 return(e_0*5*1e+3*100*(z-50e-10));      /* eF(z-z0), F in S.I. */
}
```

This particular version is set for a quantum well of width 100 Å, note how the electric field, which takes the value of 5 kVcm^{-1} in this example (in function `Hprime`), is immediately converted into S.I. units by multiplying by 10^3 (kV→V) and then by 100 (cm^{-1}→m^{-1}).

Fig. 2.3 shows the second-order correction to the ground state energy of a 100 Å infinitely deep quantum well, as calculated by the C-code above, for a series of increasing electric field strengths F. It can be seen that the change in energy is negative, which means that the application of the electric field reduces the energy of the eigenstate, and that the curve looks parabolic. Both of these effects could have been predicted from the functional form of this second-order correction, see equation (2.16). Thus the quantum-confined Stark effect gives a parabolic suppression of the energy eigenvalue with increasing electric field.

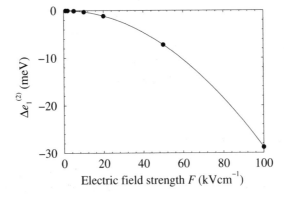

Fig. 2.3   The second-order correction $\Delta e_1^{(2)}$ to the ground state energy of a 100 Å infinitely deep quantum well when perturbed by an electric field of strength F

In the form above, the code actually prints out the value of the second-order correction after the inclusion of each additional state in the sum. This is useful because it gives an idea of how many terms are required in order to reach convergence. A quick look back at the equation for the second-order correction (equation (2.16), shows that convergence should be expected because the denominator $E_j - E_i$ increases as the states brought into the

Table 2.3 The effect of including an increasing number of states in the summation for the second-order perturbative correction

Number of states	$\Delta e_1^{(2)}$ (eV)	
	$L = 100$ Å	$L = 300$ Å
2	-0.007190	-0.582415
3	-0.007190	-0.582415
4	-0.007200	-0.583161
5	-0.007200	-0.583161
6	-0.007200	-0.583185
7	-0.007200	-0.583185
8	-0.007200	-0.583187
9	-0.007200	-0.583187
10	-0.007200	-0.583187

summation move further and further away from the state of interest. Table 2.3 illustrates this by tabulating the complete output from the same calculation above ($L = 100$ Å at a field $F = 50$ kVcm^{-1}) in comparison with a quantum well of width $L = 300$ Å.

The data in the table show that in the narrower well, where the energy levels are further apart (due to the L^{-2} dependency), the summation converges after inclusion of the terms with three higher energy states. However, in the wider well, the states are closer together in energy (hence the denominator $E_j - E_i$ is smaller), and many more terms are required before the sum converges.

2.5 TWO-DIMENSIONAL PERTURBING POTENTIALS

The one-dimensional systems that have been considered so far, serve as illustrative examples of how perturbation theory might be implemented for the calculation of energy changes in quantum systems for small changes in the potential. In fact, all of the examples above can be solved directly using the numerical shooting approach derived in Chapter 1—this is discussed in more detail and comparisons between the methods are suggested as part of the examples at the end of this chapter.

Now however, consider a two-dimensional confining potential such as that found in the cross-section of a quantum wire, as illustrated in Fig. 2.4. Any electrons in this quantum wire are free to move along the length (x-axis) of the wire but the potentials in the y-z plane localise the electron. To describe the physics of this system, the quantised electron energies arising from the confining potential would need to be calculated.

The Schrödinger equation describing the system would be:

(a)   (b)   (c)

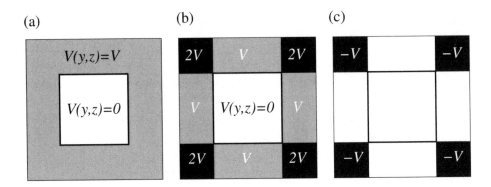

Fig. 2.4 Solving a finite-barrier quantum wire (a) by decoupling the motions (b) and applying a pertubative correction (c)

Schrödinger Equation of a Quantum Wire

$$-\frac{\hbar^2}{2m}\left(\frac{\partial^2\psi(x,y,z)}{\partial x^2} + \frac{\partial^2\psi(x,y,z)}{\partial y^2} + \frac{\partial^2\psi(x,y,z)}{\partial z^2}\right) + V(y,z)\psi(x,y,z)$$

$$= E\psi(x,y,z) \tag{2.24}$$

Now the energy E can be split into two components, one associated with the motion along the length of the quantum wire, E_x, and one associated with the motion over the cross-section $E_{y,z}$. Consider also writing the wave function as a product of components along these two mutually perpendicular directions, i.e. $\psi(x,y,z) = \psi(x)\psi(y,z)$ then:

$$-\frac{\hbar^2}{2m}\left(\frac{\partial^2\psi(x)\psi(y,z)}{\partial x^2} + \frac{\partial^2\psi(x)\psi(y,z)}{\partial y^2} + \frac{\partial^2\psi(x)\psi(y,z)}{\partial z^2}\right) + V(y,z)\psi(x)\psi(y,z)$$

$$= E_x\psi(x)\psi(y,z) + E_{y,z}\psi(x)\psi(y,z) \tag{2.25}$$

The two contributions to the motion can now be separated:

$$-\frac{\hbar^2}{2m}\frac{\partial^2\psi(x)\psi(y,z)}{\partial x^2} = E_x\psi(x)\psi(y,z) \tag{2.26}$$

$$-\frac{\hbar^2}{2m}\left(\frac{\partial^2\psi(x)\psi(y,z)}{\partial y^2} + \frac{\partial^2\psi(x)\psi(y,z)}{\partial z^2}\right) + V(y,z)\psi(x)\psi(y,z) = E_{y,z}\psi(x)\psi(y,z)$$

$$\tag{2.27}$$

and in the first of this pair of equations, no operators are acting on the y and z directions, hence $\psi(y,z)$ can be divided out, and similarly for the motion along the x axis in the second equation. Together this leaves:

Decouple Motion Along Length of Wire

$$-\frac{\hbar^2}{2m}\frac{\partial^2\psi(x)}{\partial x^2} = E_x\psi(x) \tag{2.28}$$

$$-\frac{\hbar^2}{2m}\left(\frac{\partial^2\psi(y,z)}{\partial y^2}+\frac{\partial^2\psi(y,z)}{\partial z^2}\right)+V(y,z)\psi(y,z)=E_{y,z}\psi(y,z) \qquad (2.29)$$

*Cross-
Sectional
Confinement*

The first of these two equations describes the motion of a particle along the length of the quantum wire, there is no potential and hence the solutions are just plane waves, $\psi(x)=Ae^{ikx}+Be^{-ikx}$, indicating free motion. Now consider the second equation which describes the motion over the cross-section: the wave function $\psi(y,z)$ cannot be written as a product of two components, as has been done above, because the potential, see Fig. 2.4(a), cannot be separated into a sum of one-dimensional independent potentials. Hence, although the original three-dimensional problem has been decoupled into a one-dimensional and a two-dimensional problem, no further decoupling is possible, and a full two-dimensional solution is required.

However, consider approximating the potential in Fig. 2.4(a) with that in (b). It is expected that an electron may be confined within the central ($V(y,z)=0$) region of the quantum wire, so the 'overlap' of the wave functions with the small regions of higher ($V(y,z)=2V$) potential might not be great—*it may be a perturbation*. This replacement of the potential (a) with (b) allows $V(y,z)$ to be written as a straight sum $V(y)+V(z)$, hence equation (2.29) would become:

$$-\frac{\hbar^2}{2m}\left(\frac{\partial^2\psi(y,z)}{\partial y^2}+\frac{\partial^2\psi(y,z)}{\partial z^2}\right)+(V(y)+V(z))\,\psi(y,z)=E_{y,z}\psi(y,z) \qquad (2.30)$$

*Approximate
the 2D
Potential*

This then allows the wave function $\psi(y,z)$ to be written as a product $\psi(y)\psi(z)$, giving:

$$-\frac{\hbar^2}{2m}\left(\frac{\partial^2\psi(y)\psi(z)}{\partial y^2}+\frac{\partial^2\psi(y)\psi(z)}{\partial z^2}\right)+(V(y)+V(z))\,\psi(y)\psi(z)=E_{y,z}\psi(y)\psi(z)$$

*Force
Decoupling*

$$(2.31)$$

writing $E_{y,z}$ as E_y+E_z, then:

$$-\frac{\hbar^2}{2m}\left(\psi(z)\frac{\partial^2\psi(y)}{\partial y^2}+\psi(y)\frac{\partial^2\psi(z)}{\partial z^2}\right)+\psi(z)V(y)\psi(y)+\psi(y)V(z)\psi(z)$$

$$=(E_y+E_z)\psi(y)\psi(z) \qquad (2.32)$$

which can now be decoupled to give two (easy to solve, by methods discussed in Chapter 1) one-dimensional problems by gathering terms in y and z and dividing the y-motion by $\psi(z)$ and vice versa:

$$-\frac{\hbar^2}{2m}\frac{\partial^2\psi(y)}{\partial y^2}+V(y)\psi(y)=E_y\psi(y) \qquad (2.33)$$

$$-\frac{\hbar^2}{2m}\frac{\partial^2\psi(z)}{\partial z^2}+V(z)\psi(z)=E_z\psi(z) \qquad (2.34)$$

The accuracy of the resulting energy $E_{y,z}$ can be improved by applying the perturbation given in Fig. 2.4(c), hence forcing the potential back to its true form. This perturbation is across two spatial co-ordinates (both y- and z-axes), thus the elemental volume $d\tau$ in the original definition of the first-order change in the energy becomes the elemental area $dydz$. In addition, if the quantum wire is of side L_y and L_z, and centred on the origin, then the approximated energy eigenvalue $E_{y,z} = E_y + E_z$ found from solution of equations (2.33) and (2.34) can be improved by evaluating the first-order correction:

Account for
Approximation
with
Perturbation

$$\Delta e_i = 4 \int_{L_z/2}^{\infty} \int_{L_y/2}^{\infty} \psi_i^*(y)\psi_i^*(z)(-V)\psi_i(y)\psi_i(z) \; dy \; dz \qquad (2.35)$$

where the four-fold symmetry of the perturbing potential has been used.

For further examples of this technique and practice evaluating two- and three-dimensional perturbations, see Projects 2.2 and 2.3 at the end of this chapter.

2.6 THE VARIATIONAL METHOD

Consider a quantum system described by a Hamiltonian $\mathcal{H}$ with a series of eigenstates ψ_i with corresponding energy eigenvalues E_i, then the time-independent Schrödinger equation would be:

$$\mathcal{H}\psi_i = E_i\psi_i \qquad (2.36)$$

Say the Hamiltonian of a system *was known* but the solutions themselves were not. One way to proceed might be to 'guess' a possible solution. As the eigenstates ψ_i form a complete orthonormal set (whether they are known or not) then this 'trial' wave function can be expressed as a linear combination of them:

Expand
Wave Function
as a Linear
Combination
of Basis
States

$$\psi_{\text{trial}} = \sum_{i=0}^{n} a_i\psi_i \qquad (2.37)$$

where n represents some upper limit to the number of terms that would be included in the summation *in practice*. Whatever the form of the trial wave function, assume that it can be normalised (this can be done by dividing the wave function by the square root of the integral over all space of the probability density, see Section 1.10). Mathematically, normalisation implies that:

$$\int_{\text{all space}} \psi_{\text{trial}}^* \psi_{\text{trial}} \; d\tau = 1 \qquad (2.38)$$

Given this, and substituting the form for ψ_{trial} given in equation (2.37), then:

$$\int \left(\sum_{i=0}^{n} a_i^*\psi_i^* \right) \left(\sum_{j=0}^{n} a_j\psi_j \right) \; d\tau = 1 \qquad (2.39)$$

$$\therefore \sum_{i=0}^{n} a_i^* \sum_{j=0}^{n} a_j \int \psi_i^*\psi_j \; d\tau = 1 \qquad (2.40)$$

but the eigenstates ψ_i (and ψ_j) form a complete orthonormal set, which means $\int \psi_i^* \psi_j = \langle \psi_i^* | \psi_j \rangle = \delta_{ij}$, therefore:

$$\sum_{i=0}^{n} \sum_{j=0}^{n} a_i^* a_j \delta_{ij} = 1 \tag{2.41}$$

which gives:

$$\sum_{i=0}^{n} a_i^* a_i = 1 \quad \text{or} \quad a_0^* a_0 + \sum_{i \neq 0}^{n} a_i^* a_i = 1 \tag{2.42}$$

a result that will be used later.

The energy associated with such a trial wave function (whether it be an eigenstate or not) is found by calculating the expectation value of the Hamiltonian operator[3]:

$$E_{\text{trial}} = \int \psi_{\text{trial}}^* \mathcal{H} \psi_{\text{trial}} \ d\tau \tag{2.43}$$

Energy is Expectation Value of Hamiltonian Operator

Again substituting the form for ψ_{trial} given in equation (2.37), then:

$$E_{\text{trial}} = \int \left(\sum_{i=0}^{n} a_i^* \psi_i^* \right) \mathcal{H} \left(\sum_{j=0}^{n} a_j \psi_j \right) \ d\tau \tag{2.44}$$

$$\therefore E_{\text{trial}} = \sum_{i=0}^{n} a_i^* \sum_{j=0}^{n} a_j \int \psi_i^* \mathcal{H} \psi_j \ d\tau \tag{2.45}$$

But ψ_j is an eigenstate of the Hamiltonian $\mathcal{H}$ with an energy eigenvalue E_j, i.e. $\mathcal{H} \psi_j = E_j \psi_j$, as in equation (2.36), therefore:

$$E_{\text{trial}} = \sum_{i=0}^{n} \sum_{j=0}^{n} a_i^* a_j \int \psi_i^* E_j \psi_j \ d\tau \tag{2.46}$$

Now, clearly, the constant E_j can be brought outside the integral and again using the orthonormality of the eigenstates:

$$E_{\text{trial}} = \sum_{i=0}^{n} \sum_{j=0}^{n} a_i^* a_j E_j \delta_{ij} = \sum_{i=0}^{n} a_i^* a_i E_i = a_0^* a_0 E_0 + \sum_{i \neq 0}^{n} a_i^* a_i E_i \tag{2.47}$$

where E_0 is the lowest energy eigenvalue, i.e. the ground state. Now utilising the result derived earlier in equation (2.42) then $a_0^* a_0 = 1 - \sum_{i \neq 0}^{n} a_i^* a_i$. Substituting into equation (2.47):

$$E_{\text{trial}} = \left(1 - \sum_{i \neq 0}^{n} a_i^* a_i \right) E_0 + \sum_{i \neq 0}^{n} a_i^* a_i E_i \tag{2.48}$$

---

[3]The expectation value is easier to remember in Dirac notation: $E_{\text{trial}} = \langle \psi_{\text{trial}} | \mathcal{H} | \psi_{\text{trial}} \rangle$.

$$E_{\text{trial}} = E_0 + \sum_{i \neq 0}^{n} a_i^* a_i \left(E_i - E_0 \right) \tag{2.49}$$

No matter what the particular values of a_i are, $a_i^* a_i$ is clearly greater than zero. Also for all values of i above zero, the energy eigenvalue E_i is greater than the ground state energy E_0. Thus all the terms in the summation are positive. Therefore, whatever the choice of trial wave function, whether it be similar to the ground state or completely different, when the expectation value of the energy is evaluated it will always be greater than the ground state energy, i.e.

Variational Principle

$$E_{\text{trial}} \geq E_0 \tag{2.50}$$

This is known as the 'variational principle'.

2.7 APPLICATION TO THE HYDROGEN ATOM

The variational principle is a statement of physics: whatever choice of trial wave function is made it will always yield an expectation value for the energy which is greater than the ground state energy. As far as a *computational exploitation* of the variational principle is concerned, equations (2.49) and (2.50) are useless because without knowledge of the eigenstates ψ_i of the Hamiltonian, the coefficients a_i remain unknown. The method is based around minimising the expectation value of the energy of a series of trial wave functions—with the closest approximation to the ground state being given by the trial wave function with the lowest energy. This is perhaps best illustrated by means of an example, and with this aim consider the familiar hydrogen atom, with a single negatively charged electron orbiting some distance from a nucleus which has a single positive charge, as illustrated in Fig. 2.5.

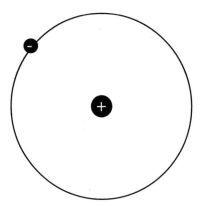

Fig. 2.5 Schematic representation of the hydrogen atom, with a single electron orbiting a positively charged nucleus

Assuming the nucleus is of infinite mass then the Hamiltonian describing such a system would consist of a kinetic energy operator and a potential energy as previously:

$$\mathcal{H} = -\frac{\hbar^2}{2m}\left(\frac{\partial^2}{\partial x^2} + \frac{\partial^2}{\partial y^2} + \frac{\partial^2}{\partial z^2}\right) - \frac{e^2}{4\pi\epsilon_0 r} \qquad (2.51)$$

where the position of the electron is given by:

$$\mathbf{r} = x\mathbf{\hat{i}} + y\mathbf{\hat{j}} + z\mathbf{\hat{k}} \qquad (2.52)$$

Following similar arguments to Section 1.12, as the wave function tends to assume the same symmetry as the potential, then the spherically symmetric potential implies that the wave function of the electron will also be spherically symmetric. Transforming the kinetic energy operator of the Hamiltonian above into spherical polar coordinates as in Appendix B, gives the Schrödinger equation for the hydrogen atom as before:

$$-\frac{\hbar^2}{2m}\left(\frac{2}{r}\frac{\partial}{\partial r} + \frac{\partial^2}{\partial r^2}\right)\psi(r) - \frac{e^2}{4\pi\epsilon_0 r}\psi(r) = E\psi(r) \qquad (2.53)$$

It was commented, when the hydrogen atom was met for the first time (in Section 1.12) that the radial wave function of the electron appeared to decay exponentially away from the nucleus. Consider choosing a trial wave function of this form:

$$\psi(r) = e^{-r/\lambda} \qquad (2.54)$$

If this form is substituted into the Schrödinger equation (equation (2.53)) then:

$$-\frac{\hbar^2}{2m}\left(\frac{2}{r}\frac{\partial}{\partial r} + \frac{\partial^2}{\partial r^2}\right)e^{-r/\lambda} - \frac{e^2}{4\pi\epsilon_0 r}e^{-r/\lambda} = E_{\text{trial}}e^{-r/\lambda} \qquad (2.55)$$

Performing the differentiation then:

$$\left[-\frac{\hbar^2}{2m}\left(-\frac{2}{\lambda r} + \frac{1}{\lambda^2}\right) - \frac{e^2}{4\pi\epsilon_0 r}\right]e^{-r/\lambda} = E_{\text{trial}}e^{-r/\lambda} \qquad (2.56)$$

This is now the simplest form for the Schrödinger equation ($\mathcal{H}\psi_{\text{trial}} = E_{\text{trial}}\psi_{\text{trial}}$) given the choice of trial wave function that has been made. The expectation value of the Hamiltonian operator, i.e. the energy E_{trial}, is then found by multiplying on the left by the complex conjugate of the trial wave function and integrating over all space (this eliminates the unknown electron position r), i.e.

$$\int \psi_{\text{trial}}^* \mathcal{H}\psi_{\text{trial}}\, d\tau = E_{\text{trial}} \int \psi_{\text{trial}}^* \psi_{\text{trial}}\, d\tau \qquad (2.57)$$

Therefore:

$$E_{\text{trial}} = \frac{\int \psi_{\text{trial}}^* \mathcal{H}\psi_{\text{trial}}\, d\tau}{\int \psi_{\text{trial}}^* \psi_{\text{trial}}\, d\tau} \qquad (2.58)$$

Note that as the wave function is not normalised, the denominator is not equal to '1', as in equation (2.43). Following this through for the hydrogen atom, then:

$$\int_0^\infty \left[-\frac{\hbar^2}{2m}\left(-\frac{2}{\lambda r} + \frac{1}{\lambda^2}\right) - \frac{e^2}{4\pi\epsilon_0 r}\right]e^{-2r/\lambda}\, dr = E_{\text{trial}} \int_0^\infty e^{-2r/\lambda}\, dr \qquad (2.59)$$

The computational implementation just involves evaluating two definite integrals; calling the one on the left-hand-side N (for numerator) and the one on the right-hand-side D (for denominator) then $E_{\text{trial}} = N/D$. Note, some caution must be exercised since two of the terms on the left-hand-side of equation (2.59) diverge (tend to large numbers) at the lower limit of integration. The algorithmic description is based around calculating these two integrals as a function of the variational parameter λ—the solution corresponds to the minimum value of E_{trial}.

Algorithm 2.3 Variational approach to ground state of hydrogen atom

$dr \Leftarrow 0.01 \times 10^{-10}$ {Define parameters}
Numerator $\Leftarrow 0$ {Initialise variables for summation}
Denominator $\Leftarrow 0$
for λ=some-lower-limit to some-upper-limit **do**
    Calculate integrals N and D using simple strip summation
    Print $E_{\text{trial}} = N/D$
end for

Variational Calculation of Hydrogen Atom

And again in C, this might appear as:

```
#include <math.h>



#define hbar 1.05459e-34
#define m 9.109534e-31
#define e_0 1.602189e-19
#define epsilon_0 8.854188e-12
#define pi 3.141593
main()
{
double   D;            /* the denominator: Etrial=N/D */
double   dr=0.01e-10;  /* the integration strip width */
double   Etrial;       /* the expectation value <psi|H|psi> */
double   lambda;       /* the variational parameter */
double   N;            /* the numerator: Etrial=N/D */
double   r;            /* radial position of electron */
```

```
for(lambda=0.01e-10;lambda<5e-10;lambda+=0.01e-10)              Vary λ
{                                        /* vary lambda */
 N=0;D=0;                        /* initialise for numerical integration */
 for(r=dr;r<10e-10;r+=dr)
 {
   N+=(-(hbar/(2*m))*hbar*(-2/(lambda*r)+1/(lambda*lambda))     Calculate N
       -e_0*e_0/(4*pi*epsilon_0*r)                            ⟨ψtrial|H|ψtrial⟩
       )*exp(-2*r/lambda)*r*r;
   D+=exp(-2*r/lambda)*r*r;                                     Calculate D
 }                                                            ⟨ψtrial|ψtrial⟩
 N*=4*pi*dr;D*=4*pi*dr;
 Etrial=N/D;                                                    Trial Energy
 printf("lambda=%f Angstrom Etrial=%f eV\n",
         lambda/(1e-10),Etrial/e_0);
}

}          /* end main */
```

Knowledge gained by the previous study of the hydrogen atom in Chapter 1 has been utilised and this is reflected in the range of values that the variational parameter is chosen to take (from 0.01 to 5 Å). Note that the integration step width dr (dr) in the program has been chosen as a first guess as 0.01 Å and further that the elemental volume is $4\pi r^2\ dr$. The problem of the divergence as $r \to 0$ is overcome simply by starting the summation very close to, but never equal to, zero. Provided that the strip width is taken small enough, this should introduce only a small error.

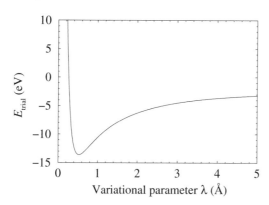

Fig. 2.6　The effect of the variational parameter λ on the expectation value E_{trial} of the Hamiltonian

Fig. 2.6 shows the results of this calculation, and plots the expectation value of the Hamiltonian, E_{trial}, as a function of the variational parameter λ. It can be seen that there is a minimum energy and inspection of the more detailed plot of the region around the minimum, in Fig. 2.7, and the raw numerical data, shows that the minimum value of the energy is -13.606 eV and occurs at $\lambda \approx 0.53$ Å.

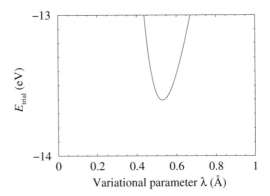

Fig. 2.7   A more detailed view of the energy minimum of Fig. 2.6

These results have been obtained with relatively coarse values for the integration strip width and λ step length. The latter in particular was chosen as 0.01 Å and clearly the accuracy of the value for λ, for which the energy minimum occurs, will never be smaller than this value. Despite these limitations the predictions are in excellent agreement with those (-13.607 eV and 0.528 Å) deduced in Section 1.12, based on the alternative method of the numerical solution of the Schrödinger equation.

2.8   APPLICATION TO BIASED QUANTUM WELLS

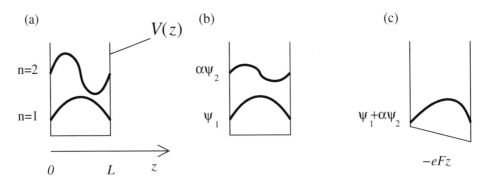

Fig. 2.8   The addition of an electric field to the ground state of an infinitely deep quantum well 'mixes' in some of the first excited state

The direct numerical solutions to Schrödinger's equation have shown that the energy of the states within a quantum well, change under the application of an electric field. The earlier approximate analysis, based around perturbation theory, showed that, for an initially symmetric system, the change in the energy of the ground state occurred only as a secondary effect. Furthermore, the perturbative approach of constructing the perturbed wave function Ψ from a linear combination of the unperturbed states ψ_i suggests an interpretation: the electric field 'mixes in' some of the higher energy states into the ground state. It was found, as displayed in Table 2.3 that *most* of the change in the ground state energy could be calculated in second-order perturbation theory, by just including the first-excited state in the series summation. This suggests that the (perturbed) ground state wave function Ψ can be written approximately as a linear combination of the unperturbed ground state wave function, plus a component (fraction) of the first-excited state wave function:

$$\Psi = \psi_{\text{trial}} \approx \psi_1 + \alpha\psi_2 \qquad (2.60) \quad \text{'Mixing'}$$

This is illustrated schematically in Fig. 2.8. The procedure is to vary the parameter α, i.e. vary the fraction of the first-excited state in the trial wave function ψ_{trial}, until the energy is minimised. As this trial wave function is not normalised, then the full expression for the expectation value of the Hamiltonian is required, this is given (as before in equation (2.58)) as:

$$E_{\text{trial}} = \frac{\int \psi_{\text{trial}}^* \mathcal{H} \psi_{\text{trial}} \, d\tau}{\int \psi_{\text{trial}}^* \psi_{\text{trial}} \, d\tau} \qquad (2.61)$$

For the infinitely deep quantum well the Hamiltonian $\mathcal{H}$ is defined only within the well $0 < z < L$. The addition of an electric field adds a potential energy component to the kinetic energy operator to give:

$$\mathcal{H} = -\frac{\hbar^2}{2m}\frac{\partial^2}{\partial z^2} - eFz \qquad (2.62)$$

Using the wave functions given in equation (2.19) then:

$$\mathcal{H}\psi_{\text{trial}} = \left(-\frac{\hbar^2}{2m}\frac{\partial^2}{\partial z^2} - eFz\right)\left[\sqrt{\frac{2}{L}}\sin\left(\frac{\pi z}{L}\right) + \alpha\sqrt{\frac{2}{L}}\sin\left(\frac{2\pi z}{L}\right)\right] \qquad (2.63)$$

$$\therefore \mathcal{H}\psi_{\text{trial}} = \left(\frac{\hbar^2}{2m}\frac{\pi^2}{L^2} - eFz\right)\left[\sqrt{\frac{2}{L}}\sin\left(\frac{\pi z}{L}\right) + 2^2\alpha\sqrt{\frac{2}{L}}\sin\left(\frac{2\pi z}{L}\right)\right] \qquad (2.64)$$

The mathematics can be taken further, or a form for the energy can be just written down, which is suitable for programming:

$$E_{\text{trial}} = \frac{\int_0^L \left[\sin\left(\frac{\pi z}{L}\right) + \alpha\sin\left(\frac{2\pi z}{L}\right)\right]\left(\frac{\hbar^2\pi^2}{2mL^2} - eFz\right)\left[\sin\left(\frac{\pi z}{L}\right) + 4\alpha\sin\left(\frac{2\pi z}{L}\right)\right]dz}{\int_0^L \left[\sin\left(\frac{\pi z}{L}\right) + \alpha\sin\left(\frac{2\pi z}{L}\right)\right]^2 dz}$$

$$(2.65)$$

where the normalisation factor $\sqrt{2/L}$ has been divided through top and bottom.

Although at first sight it might seem that, as the variational parameter α represents the fraction of first-excited state which is 'mixed' into the ground-state due to the electric field, then it should take some value between 0 and 1. However, this is not the case, the sign for α may depend on the direction of the electric field, and hence should be allowed to go negative. It is known from the previous perturbative work, from where this section is derived, that the effect of the electric field on the ground state energy is small, which in turn implies that the change in the ground state wave function is also small, therefore a sensible range over which to vary α would probably be $-1 < \alpha < 1$.

Summary

The effect of a small change (perturbation) on a system can be found by deducing the *change* to the known (whether by measurement or calculation) states (solutions) rather than resorting to another measurement or first-principles calculation. In particular, for quantum systems, the first-order change in energy of a state ψ_i due to a perturbation $\mathcal{H}'$ is given by:

$$\Delta e_i^{(1)} = \int \psi_i^* \mathcal{H}' \psi_i \ \mathrm{d}\tau$$

The second-order change in the energy is given by a sum over all the other states of the original unperturbed system:

$$\Delta e_i^{(2)} = -\sum_{j \neq i} \frac{|\int \psi_j^* \mathcal{H}' \psi_i \ \mathrm{d}\tau|^2}{e_j - e_i}$$

If the functional form of a wave function is 'guessed' then any parameters within the description can be found by minimising the expectation value of the Hamiltonian operator:

$$E_{\text{trial}} = \frac{\int \psi_{\text{trial}}^* \mathcal{H} \psi_{\text{trial}} \ \mathrm{d}\tau}{\int \psi_{\text{trial}}^* \psi_{\text{trial}} \ \mathrm{d}\tau}$$

The lowest energy obtained in this way is the best approximation to the ground state of the system.

Tasks

2.1 **Trapezium rule for evaluating definite integrals:** Section 2.3 introduced a simple strip rule for evaluating definite integrals. Adapt the algorithm (or example C-code) for calculating the first-order correction for both constant and linear perturbations to utilise the more accurate trapezium rule. Compare the results of both numerical methods for 10, 20, 50 and 100 integration strips.

2.2 **Simpson's rule for evaluating definite integrals:** Repeat Task 2.1 but this time use Simpson's rule for evaluating the definite integral numerically.

2.3 **Effect of electric field on finite-barrier quantum wells:** Use the shooting method for finite-barrier quantum wells described in Task 1.6 to generate the solutions of a series of quantum wells of different width and barrier height. Adapt the algorithm for the second-order perturbative correction to read in the wave functions generated by the shooting program and hence calculate the change in the energy of the ground and first-excited states when subject to an external field of 1, 2, 5, 10, and 20 kVcm^{-1}.

2.4 **Variational approach to hydrogen atom:** Choose a trial wave function of the form $e^{-r/\lambda}$ for an electron orbiting a single proton nucleus (assume nucleus has infinite mass), and vary the parameter λ in order to minimize the energy. Compare the wave function and the probability density with that found by the shooting method in Section 1.12.

2.5 **Variational approach to excited states of the hydrogen atom:** By choosing the unnormalised trial wave function:

$$\psi_{\text{trial}} = e^{-r/\lambda'} \left(1 - \frac{r}{\lambda'}\right)$$

deduce the value of λ' which minimises the expectation value of the Hamiltonian operator (the energy) for the hydrogen atom. What does this trial wave function represent?

2.6 **Variational approach to biased quantum wells:** Using the derivation in Section 2.8, deduce the effect of an electric field on a 100 Å wide, infinitely deep, quantum well. Compare the results with those from second-order perturbation theory as illustrated by Fig. 2.3.

Projects

2.1 **Error analysis:** The error in evaluating a definite integral numerically is clearly dependent upon the number of strips chosen for the integration, regardless of which of the three methods discussed is employed. By plotting the value of an integral evaluated numerically with 4, 8, 16, 32, 64 and 128 strips versus the number of strips, deduce the form of the function representing the error, i.e. is it proportional to the number of strips, inversely proportional, etc.

Richardson extrapolation is a technique for combining numerically obtained solutions of different accuracy in order to eliminate the error and obtain an improved result. When applied to numerically evaluated definite integrals it is referred to as Romberg integration. Deduce a formula for eliminating the error term from the strip summation and trapezium rule methods for evaluating the first-order perturbative correction. By choosing suitable examples, show how evaluating the integrals with 16 and 32 strips and combining the results, can lead to improved estimates. Compare the results with those obtained by simply increasing the number of strips in the first instance. Does either approach have an advantage?

2.2 **Two-dimensional perturbations: Application to quantum wires:** Section 2.5 discusses an approximate technique for solving finite-barrier quantum wires by forcing the decoupling of the motions perpendicular to the axis of the wire, and improving the energy estimate with the aid of perturbation theory. By using a shooting technique for the solution of finite barrier quantum wells calculate the approximate energy of the states in a series of rectangular 60×70, 100×110, 140×150, 180×190 and 220×230 Å cross-section GaAs quantum wires, surrounded by GaAlAs barriers of height $V = 100$, 200 and 300 meV (assume the electron effective mass in GaAlAs is the same as in GaAs and equal to 0.067 m_0). *Note the wires have been chosen with rectangular, rather than square, cross-sections to avoid degeneracy of these initial states. A perturbative correction for the forced decoupling in a system with degenerate states would require degenerate perturbation theory, see Chapter 3 for this.*

Write a computer program to evaluate the first-order perturbative correction to the energy level due to the removal of the potential 'pillars' of height '$-V$' along the corners of the quantum wire. Hence, deduce improved estimates for the energy eigenvalues of the quantum wires.

2.3 **Three-dimensional perturbations: Application to quantum dots:** The technique for obtaining approximate solutions to two-dimensional confining potentials as in the quantum wire of the previous Project, can be extended to the three-dimensional confining potentials of quantum 'boxes' (usually referred to as 'quantum dots'). In this case the approximate solution would be derived from the sum of the energy eigenvalues of the three independent finite-barrier quantum wells that would arise from the forced decoupling of the motions along the x-, y- and z-axes respectively. By again solving Schrödinger's equation with a shooting method applied to a finite-barrier quantum well, deduce these approximate solutions for similar sets of parameters as in the previous example. Again choose *cuboids* to avoid degenerate states.

Show that an improved estimate for the energy eigenvalues of the cubic quantum dots can be obtained by applying a perturbation of magnitude $-2V$ in cubic regions at the corners of the dot. Evaluate the first-order correction and hence obtain improved energy estimates for the energy eigenvalues of the quantum dots.

2.4 **Deduction of the perturbed wave function:** Find (from the literature) the expression for the coefficients a_j defining the contributions of the unperturbed wave functions in the expression for the *perturbed wave function*, i.e.

$$\Psi_I = \sum_{j=1}^{n} a_j \psi_j$$

Write a computer program (in the language or environment of your choice) to evaluate the Ψ in a 100 Å GaAs finite-quantum well surrounded by barriers of height 100 meV when under the influence of a variety of electric field strengths, say 1, 2, 5 and 10 kVcm^{-1} (take the electron effective mass to be equal to 0.067 m_0). Compare this approximated wave function with that obtained directly from the shooting method applied to the same system (note you will need the generalised starting conditions discussed in Project 1.2 at the end of Chapter 1).

2.5 **Two-dimensional parameter spaces:** As the electric field increases in magnitude, states higher than the first-excited are mixed into the ground state. By taking a trial wave function of the form:

$$\psi_{\text{trial}} = \psi_1 + \alpha\psi_2 + \beta\psi_3$$

explore the two-dimensional parameter space mapped by α and β for the example given in Task 2.6. Swap the component ψ_3 for ψ_4 and comment on the new outcome.

3
Matrix Methods

3.1 BASIS SETS

Consider a finite region of space with zero potential, as in Fig. 3.1. The Schrödinger equation would be of the form:

$$\mathcal{H}\psi = E\psi \quad \text{i.e.} \quad -\frac{\hbar^2}{2m}\frac{\partial^2\psi}{\partial z^2} = E\psi \qquad (3.1)$$

Schrödinger Equation for 1D Quantum Box

If the restriction is placed that the whole of the universe is encapsulated by this small region of space, then the wave functions must be entirely contained in the domain $0 \leq z \leq L$, which in turn implies that the wave functions must be zero at the boundaries at $z = 0$ and $z = L$.

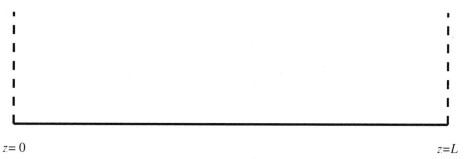

$z = 0$ $z=L$

Fig. 3.1 A one-dimensional finite region of space

Hence solutions are sought for the wave function ψ which can satisfy both the Schrödinger equation and these boundary conditions. Try $\psi(z) = A\sin kz$, where k is yet to be determined. Note that the placement of the origin at the left-hand end of the domain, rather than the centre, leads to sine terms only, this has been done deliberately as it simplifies the mathematics and the corresponding computer programs later. Substituting for ψ into equation (3.1) gives:

$$-\frac{\hbar^2}{2m}\frac{\partial^2}{\partial z^2}(A\sin kz) = E(A\sin kz) \tag{3.2}$$

$$\therefore -\frac{\hbar^2}{2m}(-Ak^2\sin kz) = E(A\sin kz) \tag{3.3}$$

which confirms that the chosen ψ is an eigenfunction of the original Schrödinger equation, equating coefficients then the energy is given by:

$$E = \frac{\hbar^2 k^2}{2m} \tag{3.4}$$

The first boundary condition ($\psi(z) = 0$ when $z = 0$) is automatically obeyed by the choice of sine terms, the second condition ($\psi(z) = 0$ when $z = L$) implies that:

$$kL = n\pi \quad \text{i.e.} \quad k = \frac{n\pi}{L} \tag{3.5}$$

where n is an integer. These solutions are illustrated in Fig. 3.2.

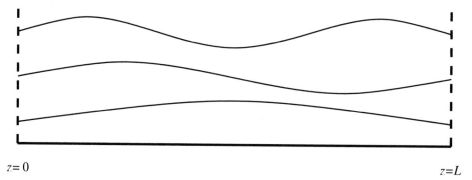

$z = 0$ $z = L$

Fig. 3.2   The three lowest energy sine function solutions in the finite region of space of Fig. 3.1

Normalising the probability distribution (see Chapter 1, equation (1.69)) gives the coefficient A as $\sqrt{2/L}$, thus the eigenstates are complete as:

Wave
Functions

$$\psi_i = \sqrt{\frac{2}{L}}\sin\left(\frac{i\pi z}{L}\right) \tag{3.6}$$

with corresponding eigenenergies obtained by substituting for k from equation (3.5) into equation (3.4):

Energy
Eigenvalues

$$E_i = \frac{\hbar^2 i^2 \pi^2}{2mL^2} \tag{3.7}$$

A set of functions ψ_i is said to be 'orthonormal' if they obey the condition:

$$\int_{-\infty}^{\infty} \psi_i^* \psi_j \ \mathrm{d}\tau = \delta_{ij} \qquad (3.8)$$

Orthonormality condition

where δ_{ij} is the Kronecker delta and is equal to 1 if $i = j$ and 0, otherwise. It can be proven that the set of solutions in equation (3.6) satisfies these conditions; consider:

$$\int_0^L \psi_i^* \psi_j \ \mathrm{d}z = \int_0^L \sqrt{\frac{2}{L}} \sin\left(\frac{i\pi z}{L}\right) \sqrt{\frac{2}{L}} \sin\left(\frac{j\pi z}{L}\right) \ \mathrm{d}z \qquad (3.9)$$

Proving Orthonormality of Quantum Box Wave Functions

Now if $i = j$ then:

$$\int_0^L \psi_i^* \psi_j \ \mathrm{d}z = \int_0^L \psi_i^* \psi_i \ \mathrm{d}z = \int_0^L \frac{2}{L} \sin^2\left(\frac{i\pi z}{L}\right) \ \mathrm{d}z \qquad (3.10)$$

Using the trigonometric identity:

$$\sin^2 A = \frac{1}{2}\left(1 - \cos 2A\right) \qquad (3.11)$$

then:

$$\int_0^L \psi_i^* \psi_i \ \mathrm{d}z = \frac{2}{L} \int_0^L \frac{1}{2}\left[1 - \cos\left(\frac{2i\pi z}{L}\right)\right] \ \mathrm{d}z \qquad (3.12)$$

$$\therefore \int_0^L \psi_i^* \psi_i \ \mathrm{d}z = \frac{1}{L}\left[z - \frac{L}{2i\pi}\sin\left(\frac{2i\pi z}{L}\right)\right]_0^L \qquad (3.13)$$

Therefore:

$$\int_0^L \psi_i^* \psi_i \ \mathrm{d}z = \frac{1}{L}\left\{L - \frac{L}{2i\pi}\sin\left(\frac{2i\pi L}{L}\right) - 0 + \frac{L}{2i\pi}\sin\left(\frac{2i\pi 0}{L}\right)\right\} = 1 \qquad (3.14)$$

Proven for $i = j$

If $i \neq j$ then using the trigonometric identity:

$$\sin A \sin B = \frac{1}{2}\left[\cos\left(A - B\right) - \cos\left(A + B\right)\right] \qquad (3.15)$$

the original integral becomes:

$$\int_0^L \psi_i^* \psi_j \ \mathrm{d}z = \frac{2}{L} \int_0^L \frac{1}{2}\left\{\cos\left[\frac{(i-j)\pi z}{L}\right] - \cos\left[\frac{(i+j)\pi z}{L}\right]\right\} \ \mathrm{d}z \qquad (3.16)$$

$$\therefore \int_0^L \psi_i^* \psi_j \ \mathrm{d}z = \frac{1}{L}\left[\frac{L}{(i-j)\pi}\sin\left[\frac{(i-j)\pi z}{L}\right] - \frac{L}{(i+j)\pi}\sin\left[\frac{(i+j)\pi z}{L}\right]\right]_0^L \qquad (3.17)$$

Both the upper and lower limits give terms which are all of the form $\sin(n\pi)$ where n is some integer. These are all equal to zero, hence:

$$\int_0^L \psi_i^* \psi_j \ \mathrm{d}z = 0, \qquad i \neq j \qquad (3.18)$$

Proven for $i \neq j$

and the orthonormality condition is proven.

The orthonormality of a set of functions also implies that any other function can be expressed as a linear combination of them[1]—*they form a 'basis set'*. Hence, if a new potential is constructed within this existing domain $0 \leq z \leq L$, then the new Schrödinger equation which arises from this has solutions which can be expressed as a linear combination of the existing orthonormal set ψ_i. This is perhaps best illustrated with an example, as in the next section.

3.2 EXPANSION

Consider the new potential $V(z)$ in Fig. 3.3, which occupies the same domain as the flat (zero) potential of the last section and whose solutions were simple sine functions.

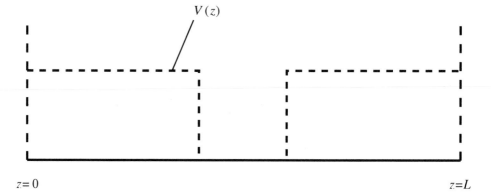

$z=0$ $z=L$

Fig. 3.3 A new potential within the existing domain of Fig. 3.1

Suppose the time-independent Schrödinger equation of the single quantum well had a solution Ψ, i.e.

$$\mathcal{H}_{\text{new}}\Psi = E\Psi \tag{3.19}$$

where this new Hamiltonian is a sum of the original Hamiltonian plus the new potential:

$$[\mathcal{H} + V(z)]\Psi = E\Psi \tag{3.20}$$

Then the work in the previous section implies that this new solution could be represented as a linear combination of the basis states ψ_j, i.e.

Series expansion

$$\Psi = \sum_{j=1}^{n} a_j \psi_j \tag{3.21}$$

---

[1]This is similar to the orthonormality of the Cartesian basis vectors $\hat{\mathbf{i}}, \hat{\mathbf{j}}$ and $\hat{\mathbf{k}}$. Clearly these unit vectors obey the orthonormality condition since, $\mathbf{i} \bullet \mathbf{i} = 1$ and $\mathbf{i} \bullet \hat{\mathbf{j}} = 0$, etc. And any vector in three-dimensional space can be represented as a linear combination of these basis vectors, i.e. $\mathbf{A} = \alpha\hat{\mathbf{i}} + \beta\hat{\mathbf{j}} + \gamma\hat{\mathbf{k}}$.

where a_j are a series of (as yet unknown) coefficients. Substituting for Ψ from equation (3.21) into equation (3.20) gives:

Substitute into Schrödinger Equation

$$[\mathcal{H} + V(z)] \sum_{j=1}^{n} a_j \psi_j = E \sum_{j=1}^{n} a_j \psi_j \qquad (3.22)$$

The aim now is to find the eigenenergies (and indeed the coefficients a_i) which characterise the solutions of the new system. With this aim, take the Hamiltonian operator within the summation (this can be done because the Hamiltonian acts on each basis function individually), then:

$$\sum_{j=1}^{n} a_j [\mathcal{H} + V(z)] \psi_j = E \sum_{j=1}^{n} a_j \psi_j \qquad (3.23)$$

$$\therefore \sum_{j=1}^{n} a_j [\mathcal{H} \psi_j + V(z) \psi_j] = E \sum_{j=1}^{n} a_j \psi_j \qquad (3.24)$$

but the basis functions are eigenfunctions of the original Hamiltonian, i.e. $\mathcal{H} \psi_j = E_j \psi_j$, therefore:

$$\sum_{j=1}^{n} a_j [E_j \psi_j + V(z) \psi_j] = E \sum_{j=1}^{n} a_j \psi_j \qquad (3.25)$$

Now multiply from the left by the complex conjugate of the lowest energy basis state, ψ_1^* and integrate over all space (in this case across the one-dimensional domain $0 \leq z \leq L$):

Multiply by ψ_1^* Integrate over all Space

$$\int_0^L \psi_1^* \sum_{j=1}^{n} a_j [E_j \psi_j + V(z) \psi_j] \ \mathrm{d}z = \int_0^L \psi_1^* E \sum_{j=1}^{n} a_j \psi_j \ \mathrm{d}z \qquad (3.26)$$

As the coefficient a_j and the energies E_j and E are constants, they can be brought out to the front of the integrals, thus giving:

$$\sum_{j=1}^{n} a_j \left[E_j \int_0^L \psi_1^* \psi_j \ \mathrm{d}z + \int_0^L \psi_1^* V(z) \psi_j \ \mathrm{d}z \right] = E \sum_{j=1}^{n} a_j \int_0^L \psi_1^* \psi_j \ \mathrm{d}z \qquad (3.27)$$

Using the orthonormality properties of the basis functions (see equation (3.8)) then:

$$\sum_{j=1}^{n} a_j \left[E_j \delta_{1,j} + \int_0^L \psi_1^* V(z) \psi_j \ \mathrm{d}z \right] = E \sum_{j=1}^{n} a_j \delta_{1,j} \qquad (3.28)$$

The object of this pursuit is to find the energies E of the new potential, but consider the last equation in more detail, i.e. write the summations explicitly:

$$a_1 \left[E_1 + \int_0^L \psi_1^* V(z) \psi_1 \ \mathrm{d}z \right] + a_2 \left[\int_0^L \psi_1^* V(z) \psi_2 \ \mathrm{d}z \right]$$

$$+ a_3 \left[\int_0^L \psi_1^* V(z) \psi_3 \ \mathrm{d}z \right] + \cdots + a_n \left[\int_0^L \psi_1^* V(z) \psi_n \ \mathrm{d}z \right] = E a_1 \qquad (3.29)$$

Not only is the energy eigenvalue E of the new potential unknown, all the coefficients a_1, a_2, a_3, etc. are also unknown and in this form the problem is insoluble.

3.3 FORMATION OF THE MATRIX EQUATION

The trick is to return to equation (3.25) and, instead of multiplying on the left by ψ_1^*, multiply on the left by ψ_2^* and then integrate over all space as before. This leads to an equation equivalent to (3.28) but with the index '1' replaced by '2':

$$\sum_{j=1}^{n} a_j \left[E_j \delta_{2,j} + \int_0^L \psi_2^* V(z) \psi_j \; dz \right] = E \sum_{j=1}^{n} a_j \delta_{2,j} \tag{3.30}$$

In a similar manner, utilizing ψ_3^* would give:

$$\sum_{j=1}^{n} a_j \left[E_j \delta_{3,j} + \int_0^L \psi_3^* V(z) \psi_j \; dz \right] = E \sum_{j=1}^{n} a_j \delta_{3,j} \tag{3.31}$$

and so on, with n equations in total. Writing the term in square brackets '[ ]' in equation (3.28) as $\mathcal{M}_{1,j}$ for short, then equation (3.29) would become:

$$\mathcal{M}_{1,1} a_1 + \mathcal{M}_{1,2} a_2 + \mathcal{M}_{1,3} a_3 + \cdots + \mathcal{M}_{1,n} a_n = E a_1 \tag{3.32}$$

and the analogous equations (3.30) and (3.31) would follow as:

$$\mathcal{M}_{2,1} a_1 + \mathcal{M}_{2,2} a_2 + \mathcal{M}_{2,3} a_3 + \cdots + \mathcal{M}_{2,n} a_n = E a_2 \tag{3.33}$$

$$\mathcal{M}_{3,1} a_1 + \mathcal{M}_{3,2} a_2 + \mathcal{M}_{3,3} a_3 + \cdots + \mathcal{M}_{3,n} a_n = E a_3 \tag{3.34}$$

with a general equation, obtained by multiplying the original equation (equation (3.25)) by ψ_i^*:

$$\mathcal{M}_{i,1} a_1 + \mathcal{M}_{i,2} a_2 + \mathcal{M}_{i,3} a_3 + \cdots + \mathcal{M}_{i,n} a_n = E a_i \tag{3.35}$$

In this shorthand form it is easier to see that this series of n equations, with n terms on the left-hand side and one term on the right-hand side of the equation, can be compacted still further and written as a matrix equation:

Matrix 'eigen' equation

$$\begin{pmatrix} \mathcal{M}_{1,1} & \mathcal{M}_{1,2} & \mathcal{M}_{1,3} & \cdots & \mathcal{M}_{1,n} \\ \mathcal{M}_{2,1} & \mathcal{M}_{2,2} & \mathcal{M}_{2,3} & \cdots & \mathcal{M}_{2,n} \\ \mathcal{M}_{3,1} & \mathcal{M}_{3,2} & \mathcal{M}_{3,3} & \cdots & \mathcal{M}_{3,n} \\ \vdots & & & & \vdots \\ \mathcal{M}_{i,1} & \mathcal{M}_{i,2} & \mathcal{M}_{i,3} & \cdots & \mathcal{M}_{i,n} \\ \vdots & & & & \vdots \\ \mathcal{M}_{n,1} & \mathcal{M}_{n,2} & \mathcal{M}_{n,3} & \cdots & \mathcal{M}_{n,n} \end{pmatrix} \begin{pmatrix} a_1 \\ a_2 \\ a_3 \\ \vdots \\ a_i \\ \vdots \\ a_n \end{pmatrix} = E \begin{pmatrix} a_1 \\ a_2 \\ a_3 \\ \vdots \\ a_i \\ \vdots \\ a_n \end{pmatrix} \tag{3.36}$$

The terms $\mathcal{M}_{i,j}$ are therefore known as 'matrix elements' and from equation (3.28), are given by:

Matrix elements

$$\mathcal{M}_{i,j} = E_j \delta_{i,j} + \int_0^L \psi_i^* V(z) \psi_j \; dz \tag{3.37}$$

3.4   SOLUTION OF THE MATRIX EQUATION

Equation (3.36) is an 'eigen' equation. The energy E is the eigenvalue and the column matrix is the eigenvector. A formal solution would proceed via the direct diagonalisation of the matrix $\mathcal{M}_{i,j}$, which would usually be achieved by recourse to a standard mathematical library, see Project 3.1. However, as this book is about *introducing* numerical techniques for the solution of common physical systems, a more 'hands on' approach will be taken.

The simplest route to a solution of the matrix equation, and a determination of the energy E, is to move the column matrix on the right-hand side of equation (3.36) over to the left-hand side to give:

$$
\begin{pmatrix}
\mathcal{M}_{1,1}-E & \mathcal{M}_{1,2} & \mathcal{M}_{1,3} & \cdots & \mathcal{M}_{1,n} \\
\mathcal{M}_{2,1} & \mathcal{M}_{2,2}-E & \mathcal{M}_{2,3} & \cdots & \mathcal{M}_{2,n} \\
\mathcal{M}_{3,1} & \mathcal{M}_{3,2} & \mathcal{M}_{3,3}-E & \cdots & \mathcal{M}_{3,n} \\
\vdots & & & & \vdots \\
\mathcal{M}_{i,1} & \mathcal{M}_{i,2} & \mathcal{M}_{i,3} & \cdots & \mathcal{M}_{i,n} \\
\vdots & & & & \vdots \\
\mathcal{M}_{n,1} & \mathcal{M}_{n,2} & \mathcal{M}_{n,3} & \cdots & \mathcal{M}_{n,n}-E
\end{pmatrix}
\begin{pmatrix}
a_1 \\ a_2 \\ a_3 \\ \vdots \\ a_i \\ \vdots \\ a_n
\end{pmatrix} = 0
\qquad (3.38)
$$

Which now represents a series of n simultaneous equations, with n unknown coefficients[2] which, if they have a solution, must have a zero determinant, i.e.

$$
\begin{vmatrix}
\mathcal{M}_{1,1}-E & \mathcal{M}_{1,2} & \mathcal{M}_{1,3} & \cdots & \mathcal{M}_{1,n} \\
\mathcal{M}_{2,1} & \mathcal{M}_{2,2}-E & \mathcal{M}_{2,3} & \cdots & \mathcal{M}_{2,n} \\
\mathcal{M}_{3,1} & \mathcal{M}_{3,2} & \mathcal{M}_{3,3}-E & \cdots & \mathcal{M}_{3,n} \\
\vdots & & & & \vdots \\
\mathcal{M}_{i,1} & \mathcal{M}_{i,2} & \mathcal{M}_{i,3} & \cdots & \mathcal{M}_{i,n} \\
\vdots & & & & \vdots \\
\mathcal{M}_{n,1} & \mathcal{M}_{n,2} & \mathcal{M}_{n,3} & \cdots & \mathcal{M}_{n,n}-E
\end{vmatrix} = 0
\qquad (3.39)
$$

Complicated though it may seem, this is just an equation with one unknown variable, i.e. it is of the form $f(E) = 0$, and hence can be solved with some of the methods introduced earlier. In particular, in Chapter 1, solutions were found to a simpler (but nonetheless similar) equation with one unknown, simply by choosing the lowest possible value of the unknown and increasing it until the function (in this case $f(E)$, say) changed sign—the value of the variable for which this occurs is then a solution. Algorithm 3.1 describes the method by which the eigenvalues E are extracted, just by inspection, by repeated calculation of the matrix determinant.

---

[2]In fact, including the energy E there are $n+1$ unknowns and only n simultaneous equations, so the system might seem insoluble. In fact, the probability condition $\int \Psi^* \Psi \, d\tau = 1$ is used to provide the additional information required. Thus the coefficients a_j are calculated to within an unknown factor and normalised according to $\sum_j a_j^* a_j = 1$.

Algorithm 3.1 Algorithmic description of matrix eigenvalue calculation

---

$\hbar \Leftarrow 1.054589 \times 10^{-34}$ {Define physical constants}

$m \Leftarrow 9.109534 \times 10^{-31}$

$L \Leftarrow 300 \times 10^{-10}$ {Define system parameters}

$n \Leftarrow 3$ {Define number of basis states}

for E=some-lower-limit to some-upper-limit **do**

  for i=1 to n **do**

    for j=1 to n **do**

      Calculate $E_j = \hbar^2 j^2 \pi^2 / (2mL^2)$

      Calculate $\mathcal{M}_{i,j}$

      Generate all matrix elements $\mathcal{M}_{i,j} - E\delta_{i,j}$

    end for

  end for

  Calculate determinant of $\mathcal{M}_{i,j} - E\delta_{i,j}$

  Print value of determinant

end for

3.5 COMPUTATIONAL IMPLEMENTATION FOR QUANTUM WELLS

The matrix approach to the solution of Schrödinger's equation by expansion in terms of some *appropriate* basis set has now been derived, and in particular, to aid visualisation of the problem, a one-dimensional 'single quantum well' potential was mentioned at the beginning of the mathematics, see Fig. 3.3. This is a useful place to start in the implementation of the method, note that the additional confining potential $V(z)$ is only going to increase the energy of the states within the original $0 < z < L$ domain, hence no eigenvalues will occur below $E = 0$ and this can be used to set the lower limit of the loop over the energy. Similarly, the upper limit can be taken as the height V_0, say, of the potential. Defining the last two parameters, the interfaces between the regions of zero and finite potential as a and b, say, completes the description of the system and allows the computer code to be written. Following the algorithmic description in the previous section, a C implementation may look similar to the following:

C Code for Matrix Solution to Energies of Finite Quantum Wells

```
#include <math.h>


#define hbar 1.05459e-34
#define m 9.109534e-31
#define e_0 1.602189e-19
#define pi 3.141593
#define L 300e-10
#define Nb 6          /* number of basis functions */
#define Nz 80         /* number of strips in integration */
```

```
main()
{
float    det();              /* routine for calculating determinant */
float    Mij();              /* the function representing Mij */
float    dE=1;               /* energy step for solution search (meV)*/
float    E;                  /* the energy E (meV) */
float    Emax=100;           /* upper limit of energy domain (meV) */
float    M[Nb][Nb][Nb];      /* the matrix Mij */
int      n=Nb;               /* number of basis states in expansion */
int      l=0;                /* the level of determinant, 0<l<n */
int      i;                  /* index over rows */
int      j;                  /* index over columns */
```

```
for(i=0;i<n;i++)             /* First set up matrix */
{
  for(j=0;j<n;j++)
  {
   M[l][i][j]=Mij(i,j);   /* Generate each matrix element */
   printf(" %f",M[l][i][j]);
  }
 printf("\n");
}
```

Set Up
Matrix $\mathcal{M}_{i,j}$
and Print Out

Loop over energy domain, E is in meV.

```
for(E=dE;E<Emax;E+=dE)
{
  for(i=0;i<n;i++)
  {
   M[l][i][i]-=dE;            /* subtract energy along diag. */
  }
 printf("%f %e\n",E,det(&l,n,M));
}
```

Subtract
Energy Along
Diagonal and
Output
Determinant
(calculated by
the function
det())

```
}  /* end main, function definitions below this line */
```

```
float
Mij(i,j)                 /* function to calculate each matrix element */
```

Function
to Calculate
Matrix
Elements

```
int      i;
int      j;
{
 float   psi();    /* the basis state */
 float   V();      /* the potential V(z) */
 float   dz;       /* the step length for integration */
 float   Ej;       /* the energy of the jth sine wave */
 float   mij=0;    /* the matrix element */
 float   z;        /* the position */
 int     iz;       /* index over z values */

 dz=L/((float)Nz);            /* calculate step length */

 /* Add kinetic energy component, note the (j+1) */

 if(i==j)mij=(hbar/m)*hbar*((float)j+1)*((float)j+1)*
            pi*pi/(2*L*L);

 /* Calculate potential integral */

 for(iz=0;iz<Nz;iz++)      /* first value of z is delta z */
 {
  z=((float)iz)*dz;        /* calculate z */

  mij+=psi(z,i)*V(z)*psi(z,j)*dz;
 }
 return(mij/(1e-3*e_0));            /* return value in meV */
}
```

Function to Calculate $\psi_i(z)$

```
float
psi(z,k)            /* calculates basis function */

float   z;
int     k;
{
 /* Note the (k+1): lowest basis state is sin(1*pi*z/L) */
 return(sqrt(2/L)*sin((float)(k+1)*pi*z/L));
}
```

Potential Function $V(z)$

```
float
V(z)                /* returns the potential V(z) */

float   z;
{
 float   lb;        /* quantum barrier width */
 float   lw=100e-10;        /* the quantum well width */
 float   V0=100*1e-3*e_0;        /* max potential */
 float   v;        /* the value of the potential */
```

```
lb=(L-lw)/2;

if((z<lb)||(z>(lb+lw)))  v=V0;
else v=0;

return(v);
}
```

Note that the energy along the diagonal is just incremented in each pass—this saves recalculating the entire matrix each time. After the creation of the matrix elements $\mathcal{M}_{ij}$, the program prints these to the standard output for diagnostic purposes, before subtracting the energy eigenvalue E from the leading diagonal and calculating the determinant. Note further, that the function for calculating the determinant is not included in this listing—it is the *first* task at the end of this chapter!

The printed matrix elements generated by the exact code listing above are as follows:

```
38.888252 0.000019 41.363693 -0.000019 -13.166617 -0.000001
0.000019 81.505409 0.000003 28.197083 -0.000021 -20.703798
41.363693 0.000003 70.427856 0.000000 20.659899 -0.000000
-0.000019 28.197083 0.000000 65.815376 0.000018 33.107010
-13.166617 -0.000021 20.659899 0.000018 82.022858 -0.000002
-0.000001 -20.703798 -0.000000 33.107010 -0.000002 81.711479
```

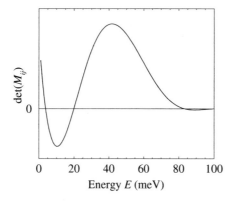

Fig. 3.4 The determinant as a function of the energy E for a single quantum well of width 100 Å and depth 100 meV, as in the code above.

Looking at such output can aid the computational process in certain circumstances. For example, in this case, see how the matrix elements between odd and even powers are (effectively) zero, i.e. for this symmetric potential:

$$\int_0^L \psi_i^* V(z)\psi_j \ \mathrm{d}z = 0, \quad \text{for } i \text{ even and } j \text{ odd, or vice versa} \qquad (3.40)$$

This knowledge can be used to reduce the computational time in future similar calculations, simply by setting these matrix elements to zero, rather than evaluating these integrals numerically.

Fig. 3.4 plots the determinant of the matrix versus the energy, as calculated by the example code above. The figure illustrates the technique well, namely that when the determinant is zero (crosses the x-axis) the energy is a solution. Examining the raw data, with an energy step of 1 meV, only allows the eigenvalues to be stated to a range, for example (in meV), $3 < E < 4$, $19 < E < 20$ and $83 < E < 84$. However, closer examination of Fig. 3.4 allows additional accuracy to be drawn from interpolating between these values and indeed 3.9, 19.8 and 83.1 are closer. As in earlier chapters, these solutions could be improved upon within the program by implementing a smaller energy step and mid-point rules, etc. However, in order determine the accuracy of the matrix approach itself, it is necessary to return to a more familiar system with simple, known, solutions—as in the next section.

3.6 CONVERGENCE TESTS: APPLICATION TO PARABOLIC POTENTIALS

Consider the potential function $V(z)$:

```
float
V(z)                        /* returns the potential V(z) */

float    z;
{
  float   v;        /* the value of the potential */
  float   z0;       /* the barrier width */

  z0=L/2;
```

Parabolic Quantum Well Potential

```
  v=((z-z0)/(100e-10))*((z-z0)/(100e-10))*0.1*e_0;

  return(v);
}
```

This represents a parabolic potential, centred in the middle of the z domain. The analysis of such potentials (see equation (1.111)) in Section 1.7 indicates that, for this example:

$$\hbar\omega = \hbar\sqrt{\frac{C}{m}} = \hbar\sqrt{\frac{2 \times 0.1 \times e}{(100 \times 10^{-10})^2 m_0}} \tag{3.41}$$

which implies $\hbar\omega = 12.345$ meV.

Consider first the effect of the number of strips in the evaluation of the integrals making up each of the matrix elements. In the program above, this is controlled by the parameter Nz. Table 3.1 shows the effect on the solutions E for this parabolic potential, using (Nb=) 4 basis functions, as a function of Nz. Note that the energy increment dE was taken as 0.1 meV for increased accuracy.

The analytical value for harmonic oscillator energy $\hbar\omega$ of 12.345 meV, implies that the ground state energy should be 12.345/2 meV$= 6.17$ meV, with the first excited state

Table 3.1   Energy eigenvalues of a parabolic potential

n	E_n (meV)		
	Nz=10	Nz=20	Nz=40
1	6.95	6.95	6.95
2	20.05	20.05	20.05
3	46.45	46.65	46.65
4	58.75	59.05	59.05

following at $3 \times 12.345/2$ meV= 18.52 meV, the second at $5 \times 12.345/2$ meV= 30.86 meV, and the third at 43.21 meV. Note that the mathematics will give only 4 solutions if only 4 basis states are included.

In this first attempt, with just 4 basis states, the calculation is executed very quickly and the resultant data, as given in Table 3.1 is in reasonable agreement with the analytical (and therefore exact) values. The table shows that, for 4 basis states, (Nz= 20) strips in the numerical integration are sufficient for convergence. In this case the most rapidly varying basis function is the fourth sine function, i.e.

$$\psi_4 = \sqrt{\frac{2}{L}} \sin\left(\frac{4\pi z}{L}\right)$$

It may be anticipated, therefore, that the more basis functions that are included, the more strips will be required in the evaluation of the integrals, because of the inclusion of more rapidly varying basis functions. As a 'rule of thumb', subsequent calculations will use 10 strips per basis function, i.e. Nz will be chosen as $10 \times$ Nb.

Fig. 3.5 shows the results of calculations of the energy eigenvalues of the same parabolic potential as above, but this time as a function of the number of basis functions. The horizontal dashed lines show the position of the analytical solutions. Note how the nth solution only appears on expansion of the basis set to n functions. Furthermore, note also, for a given number of basis functions, the lower energy solutions are reproduced better than the higher energy ones. The last set of data displayed in the figure, i.e., that corresponding to the largest basis set in this series of calculations, represents good agreement with the analytical solutions. However, it takes much effort to get to this point, and for one-dimensional potentials it is easier to implement and possible to obtain more accurate solutions, with the shooting method.

Thus, it is unlikely that such a matrix method would be used in a serious calculation for a one-dimensional system, as more *efficient* methods are available. This section has demonstrated an implementation, and in the next section, another class of problem will be introduced in which the matrix method offers the most obvious route to a solution.

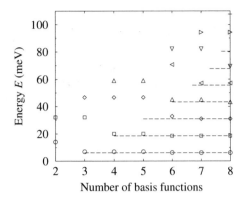

Fig. 3.5 The energy eigenvalues of the parabolic potential as a function of the number of basis states included in the expansion. The horizontal dashed lines show the position of the analytical solutions, for comparison. Note how the number of solutions produced by the method equals the number of basis functions.

3.7 SOLUTION OF SCHRÖDINGER'S EQUATION IN SEVERAL DIMENSIONS

Consider (again) a quantum wire potential, as illustrated in Fig. 3.6 and as an example of a two-dimensional system. A charge carrier is free to move along one direction (in this case the x-axis) and is confined by a finite barrier potential in the other two (y- and z-) directions.

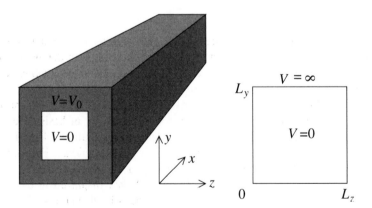

Fig. 3.6 Schematic diagram of a quantum wire and a two-dimensional quantum box

The difficult point about this system, as shown in Section 2.5, is that the confining potential $V(y, z)$ which produces the quantum mechanical effects, cannot be written as the sum $V(y) + V(z)$ and hence the motions in the y- and z- directions *do not* decouple. The Schrödinger equation, therefore, cannot be simplified beyond that given before in equation (2.29), i.e.

*Two
Dimensional
Schrödinger
Equation
for Quantum
Wire*

$$-\frac{\hbar^2}{2m}\left(\frac{\partial^2\psi(y,z)}{\partial y^2}+\frac{\partial^2\psi(y,z)}{\partial z^2}\right)+V(y,z)\psi(y,z)=E_{y,z}\psi(y,z) \quad (3.42)$$

In contrast, for the two-dimensional quantum box also illustrated in the figure, the nature of the potential, either zero inside the box or infinite outside, does mean that the motion in the y-direction can be decoupled from that in the z-direction. Hence, the infinitely deep quantum box behaves as two independent infinitely deep quantum wells, with solutions given as usual:

$$\psi_i(y)=\sqrt{\frac{2}{L_y}}\sin\left(\frac{i\pi y}{L_y}\right)\quad\text{and}\quad\psi_{i'}(z)=\sqrt{\frac{2}{L_z}}\sin\left(\frac{i'\pi z}{L_z}\right) \quad (3.43)$$

and:

$$E_y=\frac{\hbar^2 i^2\pi^2}{2mL_y^2}\quad\text{and}\quad E_z=\frac{\hbar^2 i'^2\pi^2}{2mL_z^2} \quad (3.44)$$

Such a two-dimensional system, therefore, exhibits states comprising of combinations of these independent motions. There is a different state for each unique combination of y- and z-states, hence both indices are required to specify them. The total wave function follows as the product:

$$\psi(y,z)=\psi_i(y)\psi_{i'}(z) \quad (3.45)$$

with the corresponding energy given by the sum:

$$E_{ii'}=\frac{\hbar^2\pi^2}{2m}\left(\frac{i^2}{L_y^2}+\frac{i'^2}{L_z^2}\right) \quad (3.46)$$

Now consider using these solutions as a basis set for expanding the *two-dimensional* solution $\psi(y,z)$ of the finite barrier quantum wire, i.e. let

$$\psi(y,z)=\sum_{j=1}^{n}\sum_{j'=1}^{n}a_{jj'}\psi_j(y)\psi_{j'}(z) \quad (3.47)$$

Given that there are n basis functions in each direction, then the total number of basis states in this two-dimensional problem is n^2, as opposed to n in the former one-dimensional system. In three dimensions it would be n^3, and so on.

Following the analysis in Sections 3.2 and 3.3, it can be seen that a corresponding matrix equation can be constructed which is given by:

$$\begin{pmatrix}\mathcal{M}_{11,11}&\mathcal{M}_{11,12}&\cdots&\mathcal{M}_{11,1n}&\mathcal{M}_{11,21}&\cdots&\mathcal{M}_{11,nn}\\\mathcal{M}_{12,11}&\mathcal{M}_{12,12}&\cdots&\mathcal{M}_{12,1n}&\mathcal{M}_{12,21}&\cdots&\mathcal{M}_{12,nn}\\\mathcal{M}_{13,11}&\mathcal{M}_{13,12}&\cdots&\mathcal{M}_{13,1n}&\mathcal{M}_{13,21}&\cdots&\mathcal{M}_{13,nn}\\\vdots&&&\vdots&\vdots&&\\\mathcal{M}_{ii',11}&\mathcal{M}_{ii',12}&\cdots&\mathcal{M}_{ii',1n}&\mathcal{M}_{ii',21}&\cdots&\mathcal{M}_{ii',nn}\\\vdots&&&\vdots&\vdots&&\\\mathcal{M}_{nn,11}&\mathcal{M}_{nn,12}&\cdots&\mathcal{M}_{nn,1n}&\mathcal{M}_{nn,21}&\cdots&\mathcal{M}_{nn,nn}\end{pmatrix}\begin{pmatrix}a_{11}\\a_{12}\\\vdots\\a_{1n}\\a_{21}\\\vdots\\a_{nn}\end{pmatrix}=E\begin{pmatrix}a_{11}\\a_{12}\\\vdots\\a_{1n}\\a_{21}\\\vdots\\a_{nn}\end{pmatrix}$$

Matrix Equation

$$(3.48)$$

where the general matrix element:

$$M_{ii',jj'} = E_{jj'}\delta_{ii',jj'} + \int_0^{L_z}\int_0^{L_y} \psi_i(y)\psi_{i'}(z)V(y,z)\psi_j(y)\psi_{j'}(z) \ \mathrm{d}y \ \mathrm{d}z \qquad (3.49)$$

The proof of this is the subject of Project 3.3.

The matrix, therefore, is of order n^2, which makes the problem very much more computationally demanding as there are now $n^2 \times n^2 = n^4$ matrix elements to calculate rather than just n^2 as in the one-dimensional problem. But besides this, the mathematical technique has been easily generalised to this higher dimensional problem and the computational implementation proceeds similarly. A moment of reflection brings to mind that, searching for zeroes in the determinant of the matrix $M_{ii',jj'} - E_{jj'}\delta_{ii',jj'}$, as in the one-dimensional examples, is going to be slow and that more sophisticated methods of solution, as discussed in the Tasks and Projects at the end of this chapter, will be more appropriate.

3.8 DEGENERATE PERTURBATION THEORY

In Chapter 2, perturbation theory was introduced as a means of approximating the *change* in the state of a system after a small change in the system's parameters. In particular, the case was studied of quantum systems, where a change in the potential would give rise to a change in the quantum confinement energy of an electron.

The expressions derived, see for example (2.16), have a denominator which is the difference in two energies $e_j - e_i$. Hence an immediate restriction was imposed that the effect of a perturbation could only be considered on energy levels i which were non-degenerate, because the presence of a degenerate level, j, would send the denominator to zero and hence give an infinity.

If the effect of a perturbation on a degenerate level is sought, then an alternative approach must be made—*degenerate perturbation theory*. In fact, degenerate perturbation theory is just another manifestation of the matrix methods introduced in this chapter. For example, consider the two-dimensional box, as illustrated on the right-hand side of Fig. 3.6. The energy levels of the box are given by equation (3.46) as:

$$E_{ii'} = \frac{\hbar^2\pi^2}{2m}\left(\frac{i^2}{L_y^2} + \frac{i'^2}{L_z^2}\right) \qquad (3.50)$$

and hence degeneracies clearly exist[3].

Say the system was now perturbed by some additional potential $V(y,z)$, then the energies E of the perturbed system could be deduced by expanding the perturbed wave functions $\psi(y,z)$ in terms of the wave functions of the original (unperturbed) system. This is exactly the same procedure as derived above, only from a slightly different perspective. Hence, degenerate perturbation theory is really just a particular example of a matrix method.

---

[3]For example, if $L_y = L_z$, the box is square, and E_{12} ($i = 1$, $i' = 2$) is equal to E_{21} ($i = 2$, $i' = 1$).

Summary

It has been shown that, expanding the wave function of a quantum system (described by a one-dimensional potential $V(z)$) in terms of a linear combination of basis functions ψ_i, leads to a matrix eigenequation of the form:

$$
\begin{pmatrix}
\mathcal{M}_{1,1} & \mathcal{M}_{1,2} & \mathcal{M}_{1,3} & \cdots & \mathcal{M}_{1,n} \\
\mathcal{M}_{2,1} & \mathcal{M}_{2,2} & \mathcal{M}_{2,3} & \cdots & \mathcal{M}_{2,n} \\
\mathcal{M}_{3,1} & \mathcal{M}_{3,2} & \mathcal{M}_{3,3} & \cdots & \mathcal{M}_{3,n} \\
\vdots & & & & \vdots \\
\mathcal{M}_{i,1} & \mathcal{M}_{i,2} & \mathcal{M}_{i,3} & \cdots & \mathcal{M}_{i,n} \\
\vdots & & & & \vdots \\
\mathcal{M}_{n,1} & \mathcal{M}_{n,2} & \mathcal{M}_{n,3} & \cdots & \mathcal{M}_{n,n}
\end{pmatrix}
\begin{pmatrix}
a_1 \\ a_2 \\ a_3 \\ \vdots \\ a_i \\ \vdots \\ a_n
\end{pmatrix}
= E
\begin{pmatrix}
a_1 \\ a_2 \\ a_3 \\ \vdots \\ a_i \\ \vdots \\ a_n
\end{pmatrix}
$$

where

$$
\mathcal{M}_{i,j} = E_j \delta_{i,j} + \int_0^L \psi_i^* V(z) \psi_j \ dz
$$

and E_j is the energy of the jth basis state.

Furthermore, it was shown that the above matrix equation can be rewritten in the form:

$$
\begin{pmatrix}
\mathcal{M}_{1,1} - E & \mathcal{M}_{1,2} & \mathcal{M}_{1,3} & \cdots & \mathcal{M}_{1,n} \\
\mathcal{M}_{2,1} & \mathcal{M}_{2,2} - E & \mathcal{M}_{2,3} & \cdots & \mathcal{M}_{2,n} \\
\mathcal{M}_{3,1} & \mathcal{M}_{3,2} & \mathcal{M}_{3,3} - E & \cdots & \mathcal{M}_{3,n} \\
\vdots & & & & \vdots \\
\mathcal{M}_{i,1} & \mathcal{M}_{i,2} & \mathcal{M}_{i,3} & \cdots & \mathcal{M}_{i,n} \\
\vdots & & & & \vdots \\
\mathcal{M}_{n,1} & \mathcal{M}_{n,2} & \mathcal{M}_{n,3} & \cdots & \mathcal{M}_{n,n} - E
\end{pmatrix}
\begin{pmatrix}
a_1 \\ a_2 \\ a_3 \\ \vdots \\ a_i \\ \vdots \\ a_n
\end{pmatrix}
= 0
$$

and *one* method of solution was demonstrated as searching for energy eigenvalues E which send the determinant to zero.

Considerations of *degenerate perturbation theory* illustrated that it is just a particular example of the matrix method summarised above.

Tasks

3.1 **Function for evaluating the determinant of an** $N \times N$ **matrix:** As mentioned above, the example C code in Section 3.5 for evaluating the eigenenergies from the determinant of a matrix, is complete except for the function for evaluating the determinant of the matrix. Write a function, in a programming language of your choice, to evaluate this determinant, for a general $N \times N$ matrix of real numbers. Demonstrate that it is correct by combining it with the example code in Section 3.5 and repeating the calculations for the single quantum well.

3.2 **Improved eigenvalues using the mid-point rule:** Extend the program given in Section 3.5 to give improved eigenvalues, by implementing a mid-point rule to interpolate between the energies that straddle the solution (similar to the procedure used in Section 1.8). Demonstrate its effectiveness by comparison with the solutions to the parabolic potential of Section 3.6.

3.3 **Computational time:** Repeat the calculations in Task 3.1, for a basis set with 4, 5, 6, 7 and 8 functions, and time how long it takes for the calculation to run (on UNIX-like systems this can be achieved with the 'time' command). What is the relationship between the number of basis functions and the computational time? Hence conclude whether or not this method for solution of the eigenvalues of a matrix equation is efficient.

Projects

3.1 **Direct matrix diagonalisation using standard numerical libraries or mathematical environments:** Repeat the calculations covered in Task 3.1. However, instead of finding the eigenvalues by searching for zeroes in the determinant, use an appropriate function from a numerical library such as LAPACK [13] (see http://www.lapack.org/), or by setting up the problem in a mathematical programming environment, such as Matlab, Mathcad, etc.

3.2 **Direct diagonalisation for the eigenvectors:** Extend the method employed for the solution of Project 3.1 to obtain the eigenvectors (the expansion coefficients a_i) for the few lowest energy eigenstates. Use these coefficients to construct the full solution for the wave function, and verify that they are correct by plotting them along the z-axis.

3.3 **Solution of two-dimensional potentials—the quantum wire:** Prove the form of the matrix elements in equation (3.49). Hence, adapt the example code in Section 3.5 to a two-dimensional system and calculate the effect of a finite barrier potential on the confining energy of a quantum wire. Illustrate the latter by comparing the confinement energy of an electron (of effective mass $0.067\,m_0$) localised in quantum wires, with sides in the range 40-200 Å and with potential barrier heights of 100, 200, 500, 1000 meV and ∞. In this work, retain the determinant approach for searching for the eigenvalues, but remember to restrict the number of basis states to a manageable number.

3.4 **Solution of three-dimensional, finite barrier, quantum dot potentials:** Generalise the matrix method still further to find the solutions to a three-dimensional quantum box (cuboid) with a finite barrier confining potential. As mentioned in this chapter, the number of basis states will be too large to find the eigenvalues by searching for zeroes of a determinant. Hence, implement a direct diagonalisation as discussed in Project 3.1. Verify the resulting energy eigenvalues by performing a series of calculations similar to those in Project 3.3. Are the eigenvalues' dependent on size and barrier height as expected?

4

Deterministic Simulations

4.1 COMPLEX TIME DEPENDENT SYSTEMS

So far, only static (not changing with time) systems have been considered and, in all the examples met, the physical properties of a system have been summarised by a differential equation.[1] Some of the differential equations are easy to solve, for example, the Schrödinger equation for an infinitely deep one-dimensional quantum well. Some are more complex, such as the one-dimensional parabolic potential, *one* solution for which was found by applying a standard numerical method. In these cases, and others, the system is static, there is no time evolution and the physics can be summarised, often in terms of a one-dimensional, linear differential equation.

The real world is, however, full of examples of systems that change with time, some of them are simple and the physics can be summarised in terms of a single variable equation which is solvable. Many times though, the equation describing a system is complicated and, when one of the variables is allowed to have a time dependence, it is simply too difficult to find analytical solutions. Furthermore, given the initial state of a system and the time-dependence of a variable, it is often not possible to jump straight to the solution at some time later, and it is this class of problem that is of interest in this chapter.

The way of approaching such a problem is to create a mathematical description, a 'model' of the system and to include time as the independent variable. The model then responds like the real system and the changes in the other variables, such as position, momentum, etc. can be monitored as a function of time. Such a process is called a 'simulation'—a mathematical/computational analogy of a real evolving system.

---

[1]In fact everything can be described by a differential equation, it just so happens that some of them are zeroth order!

The further qualifier, 'deterministic', in this chapter, is to indicate that the class of simulations that are to be considered here always give the same results when repeated—this is not always the case in nature, and the extended class of problems which contains random processes will be considered in the next chapter as 'stochastic simulations'.

4.2   CLASSICAL MECHANICS AND SPACE ROCKETS

A space rocket is a perfect example of a, relatively complicated, classical mechanical system—which means that it is a large macroscopic body which obeys Newton's laws of motion. The complexity arises from the fact that the mass of the rocket changes, after launch, as the fuel is burnt to produce thrust (the upwards force which lifts the rocket from the ground). In addition, the resistance of the air cannot be neglected, and both these factors must be taken into consideration if the path of the rocket is to be predicted.

Newton's Second Law states that the net force on a body is equal to the rate of change of momentum, i.e.

$$F = \frac{\mathrm{d}(mv)}{\mathrm{d}t} \tag{4.1}$$

In most systems in everyday life, the mass of the body is a constant and hence the right-hand-side of equation (4.1) reduces to:

$$\frac{\mathrm{d}(mv)}{\mathrm{d}t} = m\frac{\mathrm{d}v}{\mathrm{d}t} = ma \tag{4.2}$$

i.e. the familiar mass×acceleration. However, for a space rocket, as mentioned above, the mass is a function of time and this simplification cannot be made.

In order to deduce F, consider the forces on the rocket (for simplicity assume that the rocket follows a vertical path), as illustrated in Fig. 4.1.

Taking upwards as positive then the total force is equal to the thrust (T) minus the weight (mg) of the rocket, minus the air resistance (R), i.e.

Total Force on Rocket

$$F = T - mg - R \tag{4.3}$$

If the assumption is made that the rocket fuel is burned at a constant rate r, then the mass m, of the rocket, starts at some initial value m_0, say, and decreases linearly with time t:

Time Dependence of Mass

$$m(t) = m_0 - rt \tag{4.4}$$

Note, only those times will be considered when the rocket engines are firing, so naturally the absolute maximum time for solution will be when all the fuel is spent and the mass $m(t)$ of the rocket reaches some final value m_f. The consequence of this is that the thrust T produced will be a constant for the time domain of interest, after this, when the fuel is all used, the thrust will fall immediately to zero.

Finally, assume that the resistance to the motion from the air is proportional to the speed v of the rocket and the density ρ of air, i.e.

Air Resistance

$$R = k\rho(x)v \tag{4.5}$$

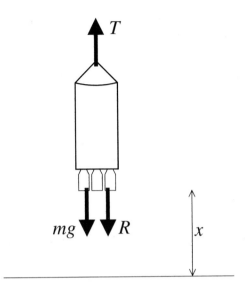

Fig. 4.1 Resolving the forces on a space rocket

where k is some proportionality constant and note, in addition, that the density of air is dependent upon the height x of the rocket above the ground. Hence, the total force F on the rocket, sometime t after launch, is given by:

$$F = T - mg - R = T - m(t)g - k\rho(x)v \qquad (4.6)$$

Substituting into equation (4.1), then:

$$T - m(t)g - k\rho(x)v = \frac{\mathrm{d}(mv)}{\mathrm{d}t} \qquad (4.7)$$

Recalling that $v = \mathrm{d}x/\mathrm{d}t$ and performing the differentiation of the product on the right-hand-side, then:

$$T - m(t)g - k\rho(x)\frac{\mathrm{d}x}{\mathrm{d}t} = \frac{\mathrm{d}m}{\mathrm{d}t}v + m(t)\frac{\mathrm{d}v}{\mathrm{d}t} \qquad (4.8)$$

which gives:

$$T - m(t)g - k\rho(x)\frac{\mathrm{d}x}{\mathrm{d}t} = \frac{\mathrm{d}m}{\mathrm{d}t}\frac{\mathrm{d}x}{\mathrm{d}t} + m(t)\frac{\mathrm{d}^2x}{\mathrm{d}t^2} \qquad (4.9)$$

Equation of Motion for a Space Rocket

Now the aim is, of course, to predict where the rocket will be (x) at any time t. Inspection of the equation of motion shows that this is far too complicated for analytical solution—the functions of time, multiplied by derivatives of time, giving it a non-linear character. However, a numerical simulation is possible and, with this aim, just rearrange equation (4.9) to give:

$$m(t)\frac{\mathrm{d}^2x}{\mathrm{d}t^2} + \left[\frac{\mathrm{d}m}{\mathrm{d}t} + k\rho(x)\right]\frac{\mathrm{d}x}{\mathrm{d}t} + m(t)g - T = 0 \qquad (4.10)$$

Consider expanding the *spatial* derivatives with respect to time in terms of finite differences: following the results from Section 1.4:

$$\frac{dx}{dt} \approx \frac{x(t+\delta t) - x(t-\delta t)}{2\delta t} \qquad \frac{d^2x}{dt^2} \approx \frac{x(t+\delta t) - 2x(t) + x(t-\delta t)}{(\delta t)^2} \tag{4.11}$$

Then equation (4.10) becomes:

*Beginnings
of a
Finite
Difference
Expansion*

$$m(t)\frac{x(t+\delta t) - 2x(t) + x(t-\delta t)}{(\delta t)^2} + \left[\frac{dm}{dt} + k\rho(x)\right]\frac{x(t+\delta t) - x(t-\delta t)}{2\delta t}$$

$$+m(t)g - T = 0 \tag{4.12}$$

Multiplying by $2(\delta t)^2$:

$$2m(t)\left[x(t+\delta t) - 2x(t) + x(t-\delta t)\right] + \delta t\left[\frac{dm}{dt} + k\rho(x)\right]\left[x(t+\delta t) - x(t-\delta t)\right]$$

$$+2(\delta t)^2\left[m(t)g - T\right] = 0 \tag{4.13}$$

Gathering terms in $x(t+\delta t)$, $x(t)$ and $x(t-\delta t)$, then:

$$\left\{2m(t) + \delta t\left[\frac{dm}{dt} + k\rho(x)\right]\right\}x(t+\delta t) - 4m(t)x(t)+$$

$$\left\{2m(t) - \delta t\left[\frac{dm}{dt} + k\rho(x)\right]\right\}x(t-\delta t) + 2(\delta t)^2\left[m(t)g - T\right] = 0 \tag{4.14}$$

Then:

$$\left\{2m(t) + \delta t\left[\frac{dm}{dt} + k\rho(x)\right]\right\}x(t+\delta t) =$$

$$4m(t)x(t) - \left\{2m(t) - \delta t\left[\frac{dm}{dt} + k\rho(x)\right]\right\}x(t-\delta t) - 2(\delta t)^2\left[m(t)g - T\right] \tag{4.15}$$

Therefore:

*Iterative
Equation
for Rocket
Trajectory*

$$x(t+\delta t) =$$

$$\frac{4m(t)x(t) - \left\{2m(t) - \delta t\left[\frac{dm}{dt} + k\rho(x)\right]\right\}x(t-\delta t) - 2(\delta t)^2\left[m(t)g - T\right]}{\left\{2m(t) + \delta t\left[\frac{dm}{dt} + k\rho(x)\right]\right\}} \tag{4.16}$$

This is an iterative equation for the position x of the space rocket—if the position is known at two times, i.e. $x(t-\delta t)$ and $x(t)$, then the position at a later time $(t+\delta t)$ can be predicted. This is very similar to the numerical solution for Schrödinger's equation, derived earlier in Chapter 1. However, it is not the same. The former contained an unknown, the energy, which was found by 'shooting forward' the solution and varying the energy until certain boundary conditions were fulfilled. In principle, all the variables on the right-hand-side of equation (4.16) are known so the path $x(t)$ can be calculated directly.

With this aim, the remaining functions have to be specified. Now the mass of the rocket as a function of time has already been discussed in equation (4.4). Given this, then:

$$\frac{dm}{dt} = -r \qquad (4.17)$$

All that remains then is the density of air as a function of height, i.e. $\rho(x)$. This has the potential for being complicated. However, for this simple computational investigation, let it be an exponentially decaying function, and furthermore assume[2] that the decay constant is 10 km. Therefore:

$$\rho(x) = \rho(0)e^{-x/10000m} \qquad (4.18)$$

In reality, this function may be more complicated, though no matter how complicated it is, it can be measured and if necessary put into the model as a list of numbers, rather than a compact mathematical expression. Note the $\rho(0)$ implies the density at sea-level ($x = 0$), take this as 1 kgm^{-3}, which is approximate, but will do here.

Now one thing that is difficult to determine is the proportionality constant k which appears in the expression for the air resistance R in equation (4.5). To do this, consider the rocket falling freely under gravity towards the ground and assume that it reaches a terminal velocity due to the resistance of the atmosphere, see Fig. 4.2.

Fig. 4.2 Using the terminal velocity to determine the air resistance

At the terminal velocity v_t, say, there are just two forces acting which are exactly equal and opposite (giving a zero resultant net force and no acceleration):

$$R = mg \qquad \therefore k\rho(x)v_t = mg \qquad (4.19)$$

---

[2]This distance is the height of Mount Everest, thus this simple function implies that the density of the atmosphere falls to e^{-1} of its sea-level value, by this height.

Hence:

$$k = \frac{mg}{\rho(x)v_{\mathrm{t}}} \tag{4.20}$$

There is considerable guesswork at this point, but say that the fully fuelled rocket, with mass $m = m_0$ would reach[3] a terminal velocity of 1,000 ms^{-1}. Then:

$$k = \frac{m_0 g}{\rho(0)v_{\mathrm{t}}} = \frac{m_0 g}{1\mathrm{kgm}^{-3} \times 1000\mathrm{ms}^{-1}} \tag{4.21}$$

All the quantities required to predict the position of the rocket for some time interval δt into the future have been specified and therefore the path can be predicted, *provided* the position of the rocket is known now and at an equivalent time interval in the past. The question then arises: *What are the initial (starting) conditions?* In the spirit of earlier chapters assume the very simplest starting conditions and allow numerical experiment to demonstrate their validity. In particular, the following choice will be made:

$$x(0) = 0 \qquad x(\delta t) = 1\mathrm{m} \tag{4.22}$$

4.3   SIMULATION OF LAUNCH TRAJECTORY

The mathematics have been derived and functional forms for all the variables defined, so progress can be made towards a computational implementation—an algorithmic description of which is given in Algorithm 4.4.1.

Algorithm 4.1 Algorithmic description of rocket simulation

$m_0 \Leftarrow 100 \times 10^3$ kg   {Define physical constants}
$g \Leftarrow 9.8$ ms^{-2} {Acceleration due to gravity}
$r \Leftarrow 1 \times 10^3$ kgs^{-1} {Rate of fuel consumption}
$\delta t \Leftarrow 1$ s   {Define computational parameters}
$x(0) \Leftarrow 0$ {Define starting values}
$x(\delta t) \Leftarrow 1$
for t=0 to 90 s **do**
    Calculate $x(t + \delta t)$
    $x(t)$ takes the value of $x(t + \delta t)$
    $x(t - \delta t)$ takes the value of $x(t)$
    Output value of $x(t + \delta t)$
end for

---

[3]To make the maths easy! Equivalent to 3,600 kmh^{-1}.

A C-code to implement this algorithm could be:

*C Code
Simulation
of Rocket
Trajectory*

```c
#include <math.h>

#define r        1e+3      /* the rate of fuel consumption kg/s */
#define m0       100e+3    /* the mass of the rocket at time t=0 */
#define g        9.8       /* acceleration due to gravity */

main()
{
float    m();             /* the rocket mass at time t */
float    p();             /* density of air */
float    k;               /* air resistance constant */
float    dt=1;            /* the time increment */
float    t;               /* the time */
float    T;               /* rocket engine thrust */
float    x0,x1,x2;        /* the position at time t-dt,t and t+dt */
```

*Define
Thrust*

```c
T=1.1*m0*g;               /* make thrust > initial weight */
```

*Define
Air
Resistance
Constant*

```c
k=m0*g/1000;              /* see notes */

x0=0;                     /* set starting conditions, x(0)=0 */
x1=1;                     /* x(dt)=1 m */

printf("0.000000 %f\n",x0);      /* and print them out */
printf("%f %f\n",dt,x1);
```

*Loop
Over
Time*

```c
for(t=dt;t<90;t+=dt)      /* start time loop */
{
 x2=(4*m(t)*x1-(2*m(t)-dt*(-r+k*p(x1)))*x0-2*dt*dt*(m(t)*g-T))/
    (2*m(t)+dt*(-r+k*p(x1)));
 x0=x1;
 x1=x2;

 printf("%f %f\n",t+dt,x2);
}

}               /* end main */
```

Function definitions below this line

Mass as a Function of Time

```
float
m(t)                            /* the mass of the rocket */

float    t;                     /* the time */
{
  return(m0-r*t);
}
```

Density of Air

```
float
p(x)

float    x;                     /* the height of the rocket */
{
  return(exp(-x/10000));
}
```

Note, in this simulation, the initial mass m_0 of the rocket has been taken as 100 tonnes, of which 90 tonnes is fuel, say. When coupled with the rate of fuel consumption as 1 tonne per second, then the maximum time for the simulation in this form is 90 s. Furthermore, note the approximation of taking the force due to gravity as a constant, this is a reasonable approximation for vehicles close to the Earth, in fact replacing:

$$mg \quad \text{with} \quad \frac{GM_\mathrm{E}m}{(R_\mathrm{E} + x)^2} \tag{4.23}$$

where M_E and R_E are the mass and radius of the Earth respectively, is more accurate and could be easily implemented. The thrust, T, has been taken as 1.1 times the initial weight of the rocket, this ensures that the rocket can lift off the launch pad.

4.4   VERIFICATION OF COMPUTATIONAL PARAMETERS

In numerical simulations, it is important that the results are insensitive to the particular values chosen for the numerical parameters, i.e. things like the time interval δt, as opposed to the set of parameters which describe the physics, such as the mass of the rocket, the acceleration due to gravity and the decay constant for the density of air with height.

Fig. 4.3 shows the effect of implementing the starting conditions for launch as listed in equation (4.22), in comparison with the starting conditions $x(0) = 0$ and $x(\delta t) = 0$. The physical interpretation of these are that, in the first instance, the rocket is stationary on the launch pad at time $t = 0$ and then at a time δt later (taken as 1s in these simulations) it is at a height of 1 m. In the latter case, the starting conditions say that the rocket is still on the launch pad δt later—despite this, the forces within the simulation take over and predict identical paths for the remainder of the trajectory, as illustrated in the figure.

The second (and final) computational parameter is the time interval δt from which the first and second derivatives of the height (i.e. the speed and acceleration) were

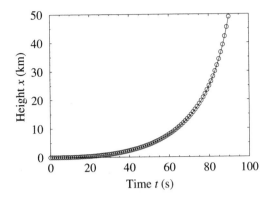

Fig. 4.3 The effect of different initial conditions (solid line: $x(0) = 0$ m, $x(\delta t) = 0$ m, circles: $x(0) = 0$ m, $x(\delta t) = 1$ m)

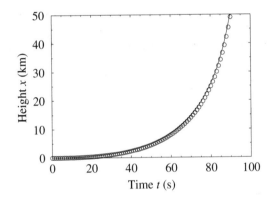

Fig. 4.4 The effect of different time intervals (solid line:$\delta t = 0.1$ s, circles:$\delta t = 1$ s)

approximated. Clearly the predictions of the simulation should be independent of the value chosen for δt. Fig. 4.4 illustrates the results of calculations with δt chosen as 0.1 s and δt chosen as 1 s. The predicted trajectories are identical for the first half of the 'engine burn', but towards the end of the simulation there is a small discrepancy. However this is small when considering that the parameter has been changed by a factor of 10. Further simulations with other values of δt confirm that values similar to those used, produce stable predictions, which serves as evidence in support of this model. This latter point is very similar to the scenario relating to the spatial interval δz in the shooting-technique approach to the solution of Schrödinger's equation in Chapter 1.

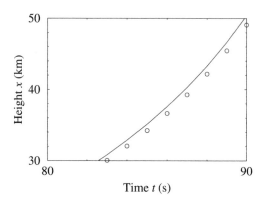

Fig. 4.5 The effect of different ground-level air densities (solid line: $\rho(0) = 0.9$ kgm^{-3}, circles: $\rho(0) = 1$ kgm^{-3})

4.5 FUN WITH SIMULATIONS

The point of building a simulation is to be able to perform numerical experiments on a computer to explore the parameter space of the physical system that it models. For example, it is extremely useful to investigate the effects of increased initial mass of a rocket and the rate and duration for which fuel is burnt, rather than deducing these empirically by performing repeated actual experiments—*rockets are extremely expensive!* This is therefore the point at which 'fun' can be had.

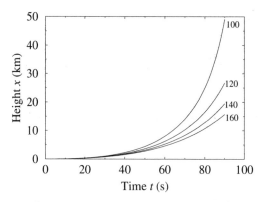

Fig. 4.6 The effect of different initial rocket masses (given in tonnes)

Say, for example, that an extremely low atmospheric pressure exists over the rocket launch pad, such that the air density at ground level is only 90% of its average value—*what would be the consequences?* Fig. 4.5 simulates such a scenario in comparison with the usual conditions. The data show that the effect is in fact quite small, and the plot needs to be focussed just on the high altitude portion of the trajectory, near the end of the engine burn, to show the difference in height between the trajectories of about one km. So it would be safe to launch, provided the winds usually associated with the low pressure don't produce too large a lateral force.

As a further example of what is possible, Fig. 4.6 displays the results of simulations with initial rocket masses greater than the 100 tonnes in the previous examples. The rocket is able to lift from the ground as before because the thrust remains as 1.1 times the initial weight of the rocket. Observe though, how the increased work that the rocket has to do against the greater gravitational force, reduces the height that it can reach. The fuel would have to be burnt at a greater rate or it would have to be burnt for longer to achieve the same altitude.

There are more examples of interesting features that can arise from the simulation in the Tasks and Projects at the end of this chapter.

4.6 INTRODUCTION TO DIFFUSION

All substances attempt to move from areas where they are present in high concentrations to areas of low concentration—a process which is known as 'diffusion'. For example, obvious though it seems, if the curve $c(z)$ in Fig. 4.7 represented the concentration of water in a trough, then the water would fall very rapidly from the region of high concentration to the region of low! Crude though it may seem, this is an example of diffusion. It also occurs in gaseous systems, for example, as smoke gradually disperses in an enclosed room; and it is an important mechanism in solids. In the context of semiconductors, and in particular semiconductor heterostructures, which are being increasingly incorporated into modern optoelectronic devices, it is clear that diffusion of *material species* could be important as their very nature derives from discontinuous changes in materials. Fig. 4.7 could therefore represent an n- or p-type dopant, in either a bulk semiconductor, i.e. a 'homojunction', or an alloy component, e.g. Al at a GaAs/Ga$_{1-x}$Al$_x$As 'heterojunction'.

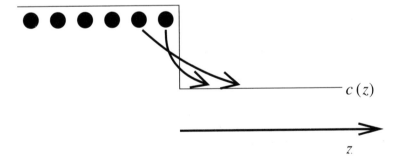

Fig. 4.7 Simple illustration of diffusion; a diffusant moves from an area of high to low concentration

Diffusion at such boundaries is a strong possibility, particularly during the elevated temperatures often used during their fabrication, or the lower but more prolonged heating that may occur during normal device operation. Any movement of material, e.g. Al at a GaAs/Ga$_{1-x}$Al$_x$As heterojunction will 'blur' the interface, i.e. the change from one material to the other will occur via a range of intermediatary alloys. Such a process is known as 'interface mixing' and is represented schematically in Fig. 4.8. The change in profile of the junction will inevitably alter the electronic properties of the system, which will in turn affect the device characteristics and therefore the operating lifetime. The motivations behind modelling diffusion in semiconductor systems is to increase understanding, in order to be able to control or prevent it, design a device in which its impact will be minimised, or predict the lifetime of a device.

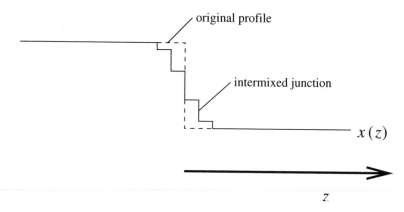

Fig. 4.8 The total amount of diffusant, represented by the area under the curve, remains constant in an intermixing heterojunction

A mathematical description of diffusion was first derived by Fick in 1855, who summarised it in two laws. The first of these stated that the steady state flux through a plane is proportional to, but in the opposite direction to, the concentration gradient, i.e.

Ficks
First Law
of Diffusion

$$\text{flux} = -\mathcal{D}\nabla c \tag{4.24}$$

where the spatial derivative ∇ is given by equation (1.37) and the constant of proportionality $\mathcal{D}$ has become known as the 'diffusion coefficient'. This law and those that derive from it are 'macroscopic' laws in that they attempt to describe the distribution of a substance on a scale which is much larger than the particles (usually atoms or molecules) that make up the substance. The diffusion coefficient $\mathcal{D}$ describes the average rate at which particles making up the diffusing substance, move. In this work the mechanisms by which the diffusion proceeds are not of interest. On a microscopic scale, some of the particles will sometimes move more than the average and others less. The diffusion coefficient will depend on the materials present in the system of interest and it may be constant, or it may have a more complicated functional form, for example:

- $\mathcal{D} = D_0$, a constant, for simple diffusion problems.

- $\mathcal{D} = D(c)$, a function of the concentration as encountered in non-linear diffusion problems. Note, that as $c = c(z)$ then $\mathcal{D}$ is intrinsically a function of position too.

- $\mathcal{D} = D(z)$, a function of position only. Such a situation can occur in ion implantation[4] problems, where the energetic ions induce lattice damage into a crystal. The vacancies created during such a process can induce 'vacancy-assisted diffusion'. The diffusion coefficient may be linearly dependent on the vacancy concentration, which itself is dependent on the depth into the material, hence the diffusion coefficient is dependent on position (depth).

- $\mathcal{D} = D(t)$, a function of time, as could occur during the annealing of radiation damage. For example, after a material has been ion implanted, the lattice damage can be repaired by annealing. As the lattice is repaired the number of vacancies (and clearly the number of interstitials) decreases, thus reducing the diffusion coefficient describing vacancy-assisted diffusion.

Atomistic models of diffusion *do* exist and are generally centred around stochastic simulations (see the next chapter) of the movement of individual atoms.

The second law stated that the time dependency of the concentration at any given point is given by the divergence of the flux:

$$\frac{\partial c}{\partial t} = -\nabla \cdot \text{flux} \tag{4.25}$$

Ficks Second Law of Diffusion

Using equation (4.24) to substitute for the flux in equation (4.25) leads to the 'diffusion equation':

$$\frac{\partial c}{\partial t} = \nabla \cdot (\mathcal{D}\nabla c) \tag{4.26}$$

Diffusion Equation

Many everyday problems, such as diffusion at an interface, can be described in one dimension (one spatial coordinate), in which case the diffusion equation above simplifies to[5]:

$$\frac{\partial c}{\partial t} = \frac{\partial}{\partial z} \mathcal{D} \frac{\partial c}{\partial z} \tag{4.27}$$

The aim in finding solutions to the diffusion equation is to be able to predict, given an initial diffusant concentration $c(z)$, the concentration at any position z any time t later. Now there are many particular solutions for specific initial diffusant distributions and for a limited set of functional forms of the diffusion coefficient[6]. However, each situation is considered separately and a new numerical solution is sought. It would be advantageous if a general method could be developed which was applicable to *all* situations regardless of the initial conditions and the functional form of the diffusion coefficient. In fact, it will be shown in subsequent sections that the diffusion equation is another ideal system in which to develop a deterministic simulation—the generation of a mathematical/computational model which mirrors the behaviour of the real physical system and allows for predictions of the concentration of the diffusant at time intervals into the future.

---

[4]A process by which a material is bombarded by energetic ions from an accelerator.

[5]For further discussion on these laws see W. D. Callister, *Materials Science and Engineering*, (Wiley, New York, 1985) p. 66.

[6]See, for example J. Crank, *The Mathematics of Diffusion*, (Oxford, London, 1956).

4.7 THEORY

As mentioned in the previous section, the *most general* one-dimensional diffusion equation for a diffusant distribution represented by $c(z)$ is given by Fick's second law:

One-Dimensional Diffusion Equation

$$\frac{\partial c}{\partial t} = \frac{\partial}{\partial z}\left(\mathcal{D}\frac{\partial c}{\partial z}\right) \qquad (4.28)$$

where the diffusion coefficient $\mathcal{D}$ could have temporal t, spatial z and concentration c dependencies, i.e. $\mathcal{D} = \mathcal{D}(c, z, t)$. Given this, then the derivative with respect to z operates on both factors, resulting in the following:

$$\frac{\partial c}{\partial t} = \frac{\partial \mathcal{D}}{\partial z}\frac{\partial c}{\partial z} + \mathcal{D}\frac{\partial^2 c}{\partial z^2} \qquad (4.29)$$

which is a second-order (as determined by the second derivative $\partial^2/\partial z^2$), non-linear (given by the $(\partial/\partial z)^2$) differential equation.

Learning from the benefits of expanding the derivatives, in both the Schrödinger equation and the equations of motion earlier in this chapter, with finite differences, would then appear to offer a promising way forward.

Recall the finite difference approximations to first and second derivatives, i.e.

Finite Difference Approximations

$$\frac{\partial f}{\partial z} \approx \frac{f(z + \delta z) - f(z - \delta z)}{2\delta z} \qquad \frac{\partial^2 f}{\partial z^2} \approx \frac{f(z + \delta z) - 2f(z) + f(z - \delta z)}{(\delta z)^2} \qquad (4.30)$$

Then the *spatial* derivatives in equation (4.29) can be expanded giving the diffusion equation in the form:

Expand the Spatial Derivative

$$\frac{dc}{dt} = \left[\frac{\mathcal{D}(c, z + \delta z, t) - \mathcal{D}(c, z - \delta z, t)}{2\delta z}\right] \times \left[\frac{c(z + \delta z, t) - c(z - \delta z, t)}{2\delta z}\right]$$

$$+\mathcal{D}(c, z, t)\left[\frac{c(z + \delta z, t) - 2c(z, t) + c(z - \delta z, t)}{(\delta z)^2}\right] \qquad (4.31)$$

All that remains is the derivative with respect to time. Following the principle of simplicity, *in this case* a more simple forward difference is going to be employed rather than the more usual central difference, i.e. approximate:

Consider the Time Derivative

$$\frac{dc}{dt} \text{ as } \frac{c(z, t + \delta t) - c(z, t)}{\delta t} \text{ rather than } \frac{c(z, t + \delta t) - c(z, t - \delta t)}{2\delta t} \qquad (4.32)$$

as would be expected from the expansion in equation (4.30). Substituting for $\partial c/\partial t$ into equation (4.31) finally gives:

Iterative Equation for Diffusion

$$\frac{c(z, t + \delta t) - c(z, t)}{\delta t} = \left[\frac{\mathcal{D}(c, z + \delta z, t) - \mathcal{D}(c, z - \delta z, t)}{2\delta z}\right]$$

$$\times \left[\frac{c(z + \delta z, t) - c(z - \delta z, t)}{2\delta z}\right]$$

$$+\mathcal{D}(c,z,t)\left[\frac{c(z+\delta z,t)-2c(z,t)+c(z-\delta z,t)}{(\delta z)^2}\right] \qquad (4.33)$$

Thus if the function $c(z,t)$ is known (at all positions z) when $t=0$, i.e. it is simply the initial profile of the diffusant, and the diffusion coefficient $\mathcal{D}$ is fully prescribed, then it is apparent from equation (4.33) that the concentration c at any point z can be calculated a short time interval δt into the future.

4.8 BOUNDARY CONDITIONS

Thus, given the initial diffusant profile and a fully prescribed diffusion coefficient, everything is in place for predicting the profile of the diffusant at any time in the future, i.e. $c(z,t+\delta t)$, except for the conditions at the ends of the system. These cannot be calculated with the iterative equation (equation (4.33)), as the equation requires points which lie either side of z, which for the end points would lie outside the z-domain.

For diffusion from an infinite source, it may be appropriate to fix the diffusant concentration c at the two end points, e.g. $c(z=0,t)=c(z=0,t=0)$. Alternatively, the concentrations c at the limits of the z-domain could be set equal to the adjacent points which can be deduced from equation (4.33). Physically this defines the diffusing structure as a closed system, with the total amount of diffusant remaining the same. It is these latter 'closed system' boundary conditions which will be employed in the following examples.

4.9 NUMERICAL IMPLEMENTATION

Algorithm 4.2 Algorithmic description of diffusion simulation

---

$D_0 \Leftarrow 0.01\ \mathrm{m^2 s^{-1}}$	{Define diffusion coefficient}
$L \Leftarrow 1.0\ \mathrm{m}$	{Define total length of system}
$N \Leftarrow 100$	{Define number of z points}
$T \Leftarrow 100\ \mathrm{s}$	{Define total time for diffusion}
Calculate δt as a fraction of $t_{\mathcal{D}}$	
Define initial diffusant profile $c(z)$	
for $t = \delta t$ to T **do**	
for $z = \delta z$ to $L - \delta z$ **do**	
Calculate each new value of $c(z,t+\delta t)$	
end for	
$c(0,t+\delta t) \Leftarrow c(\delta z,t+\delta t)$	{Impose boundary conditions}
$c(L,t+\delta t) \Leftarrow c(L-\delta z,t+\delta t)$	{in this case—closed system}
Copy new concentration to old	
end for	
Output final concentration $c(z,T)$	

Again it is useful to deconstruct the mathematics of the iterative solution of the diffusion equation into a form more akin to a computational implementation. One example of such a procedure is given in Algorithm 4.2.

Again using C as an example programming language to illustrate a specific implementation of this algorithm, could give a code such as:

```c
#include <math.h>

#define D0 0.01          /* define baseline diffusion coefficient*/
#define L 1.0            /* define the total length of system */
#define N 100            /* define number of spatial points */

main()
{
float   D();            /* the diffusion coefficient */
float   c[N];           /* array containing c(z,t) */
float   C[N];           /* array containing c(z,t+dt) */
float   dt;             /* the time increment */
float   dz=L/N;         /* the distance between z points */
float   t;              /* the time */
float   T=0.01;         /* the time of diffusion */
float   z;              /* the spatial coordinate */
int     iz;             /* index over spatial coordinate z */
```

Define δt

```c
dt=(L*L/D0)/100000;
```

Define Initial Profile

```c
for(iz=0;iz<N;iz++)       /* define initial diffusant profile */
{
 if(iz<N/2)c[iz]=1;
 else c[iz]=0;
}
```

Loop over Time

```c
for(t=dt;t<=T;t+=dt)      /* start time loop */
{
 for(iz=1;iz<N-1;iz++)    /* calculate new concentration at each position */
 {
  z=iz*dz;                /* calculate the z-coordinate */
```

Calculate New Concentration at each z

```c
  C[iz]=dt*(
            (D(c[iz],z+dz,t)-D(c[iz],z-dz,t))/(2*dz)
            *(c[iz+1]-c[iz-1])/(2*dz)
            +D(c[iz],z,t)*(c[iz+1]-2*c[iz]+c[iz-1])/(dz*dz)
           )
         +c[iz];
 }
 C[0]=C[1];C[N-1]=C[N-2];      /* impose 'closed system' boundaries */
 for(iz=0;iz<N;iz++)c[iz]=C[iz];     /* old concentration->new */
}
```

```
for(iz=0;iz<N;iz++)        /* output c(z,T), final diffusant profile */
  printf("%f %f\n",iz*dz,c[iz]);

}           /* end main */
```

Function definitions below this line

*Function
Defining
$\mathcal{D}$*

```
float
D(c,z,t)                   /* the diffusion coefficient */

float    c;                /* the concentration at z and t */
float    z;                /* the position */
float    t;                /* the time */
{
  return(D0);
}
```

4.10 CONVERGENCE TESTS

Figure 4.9 shows the time evolution of the concentration profile of a diffusant, initially localised entirely in the left half of the system, i.e.

$$c(z, t = 0) = 1, \quad 0 < z < 0.5\text{m} \qquad c(z, t = 0) = 0, \quad 0.5\text{m} < z < 1\text{m} \qquad (4.34)$$

using the closed-system boundary conditions as described above, and the particular numerical implementation illustrated in the previous section. Clearly the 'closed' nature of the system can be seen—the total amount of diffusant remains the same, and ultimately as would be expected for the a system such as the 'water step', the concentration reaches a constant value. In the case of water, this could be looked upon as minimising the potential energy.

The time interval δt for the iteration was chosen by considering the 'characteristic diffusion time' given by combining the diffusion coefficient with the total length L of the structure:

$$t_{\mathcal{D}} = \frac{L^2}{\mathcal{D}} \qquad (4.35)$$

*Characteristic
Diffusion Time*

This quantity represents the approximate time it will take for the system to diffuse completely and reach equilibrium. In the above example, the total length of the system was 1 m and the diffusion coefficient was taken as a constant, D_0, with the value 0.01 m^{2}s^{-1}, hence:

$$t_{\mathcal{D}} = \frac{L^2}{D_0} = \frac{(1\text{m})^2}{0.01\text{m}^2\text{s}^{-1}} = 100\text{s} \qquad (4.36)$$

which is exactly as found in the simulations illustrated in Fig. 4.9. It was found by numerical experiment that taking δt as $\frac{1}{100,000}$th of this value (i.e. 0.001 s) produced stable solutions—larger values produced instabilities and physically incorrect results as illustrated by negative concentrations—smaller values just gave increased computational time.

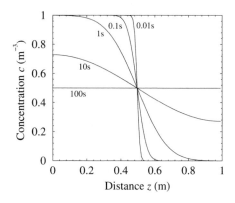

Fig. 4.9 Time evolution of the concentration c of a diffusant initially contained entirely in the left-half of the system, the diffusion coefficient was taken as a constant and equal to 0.01 m^2s^{-1}

4.11 CONSTANT DIFFUSION COEFFICIENTS AND UNIVERSALITY

In the preceding section, a constant diffusion coefficient was used in an example to illustrate the simulation. The form of the $\mathcal{D}$ was not relevant, many of the features of the model could have been demonstrated with more complicated functional forms. But, in fact, if the diffusion process can be described by a constant diffusion coefficient, $\mathcal{D} = D_0$, then the general diffusion equation, equation (4.28), and its equivalent computation form, (equation (4.33)) simplifies to the following:

Constant
Diffusion
Coefficient

$$\frac{\partial x}{\partial t} = D_0 \frac{\partial^2 x}{\partial z^2} \tag{4.37}$$

which has error-function solutions[7] for the case of diffusion at the interface ($z = 0$) of a semi-infinite slab of concentration c_0, i.e.

Error
Function
Solutions

$$c(z) = \frac{c_0}{2}\text{erfc}\frac{z}{2\sqrt{D_0 t}} \tag{4.38}$$

Universality

where 'erfc' is the complementary error function[8]. The point is that the diffusion coefficient and the time appear as a product and hence the prediction is that the solutions must be dependent only on $D_0 t$, i.e. they are universal, and this offers another opportunity to test the newly derived numerical simulation.

The continuous line in Fig. 4.10 displays the results of diffusing the same 'step' system as in the previous examples, for 20 s with $D_0 = 0.01$ m^2s^{-1}. In comparison with this, the circular symbols are the results of diffusing the same initial system for just 2 s but with $D_0 = 0.1$ m^2s^{-1} (note only every fifth symbol is plotted, $\delta z = 0.01$ m). Clearly the numerical solution *does* reproduce the universality as well—a point which is not at all obvious from the numerical form of the diffusion equation in equation (4.33).

---

[7] See J. Crank, *The Mathematics of Diffusion*, (Oxford, London, 1956).
[8] See M. Abramowitz and I. A. Stegun *Handbook of Mathematical Functions*, (Dover, New York, 1965) p. 295.

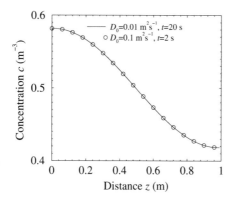

Fig. 4.10 The universality of the solutions for constant diffusion coefficients

4.12 NON-LINEAR DIFFUSION

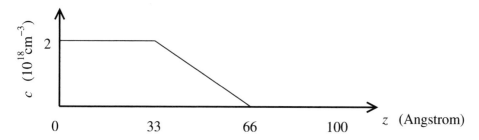

Fig. 4.11 The initial diffusant profile, note the total length of the system is 100 Å

Just to illustrate the simplicity and power of this numerical method for simulating diffusion problems, consider the rather complex example system where the diffusion coefficient is dependent upon the concentration of the diffusant and the initial diffusant profile is not a simple step function. In particular, consider diffusion of a dopant species at a linearly graded homojunction as illustrated in Fig. 4.11. Choose:

$$\mathcal{D} = D_0 \exp\left(\frac{c}{c_0}\right) \tag{4.39}$$

where D_0 is a constant and equal to $1\ \text{Å}^2\text{s}^{-1}$ and c_0 is equal to $1\times10^{18}\ \text{cm}^{-3}$.

The characteristic diffusion time is then:

$$t_\mathcal{D} = \frac{L^2}{\mathcal{D}} = \frac{(100\text{Å})^2}{1\text{Å}^2\text{s}^{-1}} = 10,000\text{s} \tag{4.40}$$

where the $\mathcal{D}$ has just been taken as the constant multiplier D_0 from the expression in equation (4.39) to give a 'ball park' figure. This choice, and the subsequent choice of the time interval δt as $\frac{1}{100,000}$ th of this (as found in Section 4.11) is borne out by the results of the simulations, displayed in Fig. 4.12. The latter are stable and indeed show that, after the characteristic diffusion time of 10,000 s has passed, the system has completely diffused.

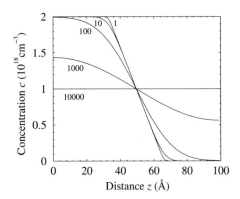

Fig. 4.12 The diffusant profile after diffusion times of 1, 10, 100, 1000 and 10,000 s

See the literature for further applications of this numerical method[9].

[9]P. Harrison, *Quantum Wells, Wires and Dots*, (Wiley, Chichester, 1999).

Summary

Generalised equations of motion for classical particles:

$$f\left(\frac{d^2x}{dt^2}, \frac{dx}{dt}, x, t\right) = 0$$

can be solved (i.e. the function $x(t)$ can be found) by expanding the first- and second-order derivatives in terms of their finite difference approximations:

$$\frac{dx}{dt} = \frac{x(t + \delta t) - x(t - \delta t)}{\delta t} \qquad \frac{d^2x}{dt^2} = \frac{x(t + \delta t) - 2x(t) + x(t - \delta t)}{(\delta t)^2}$$

and forming an iterative equation of the form:

$$x(t + \delta t) = g\left[x(t), x(t - \delta t), t\right]$$

This allows the position of the particle to be predicted a time δt into the future *provided* it is known at the present time and a time δt previously.
The diffusion equation:

$$\frac{\partial c}{\partial t} = \nabla \cdot (\mathcal{D}\nabla c)$$

was considered as another example problem for a deterministic simulation. It was shown that an iterative equation could be generated by expanding the spatial derivatives with central finite differences and the time derivative with a forward derivative, i.e.

$$\frac{c(z, t + \delta t) - c(z, t)}{\delta t} = \left[\frac{\mathcal{D}(c, z + \delta z, t) - \mathcal{D}(c, z - \delta z, t)}{2\delta z}\right]$$

$$\times \left[\frac{c(z + \delta z, t) - c(z - \delta z, t)}{2\delta z}\right]$$

$$+ \mathcal{D}(c, z, t) \left[\frac{c(z + \delta z, t) - 2c(z, t) + c(z - \delta z, t)}{(\delta z)^2}\right]$$

This allowed the concentration c of the diffusant to be predicted a time interval δt into the future *provided it was known for all positions at the current time.*

Tasks

4.1 **Launch with reduced thrust:** Find, by numerical experiment, what happens (within the constraints of this model) when the thrust is less than the initial weight of the rocket. How should the computer program be improved to account for this?

4.2 **Rate of fuel consumption:** Extend the rocket trajectory simulation to make the thrust proportional to the rate of fuel consumption. Given this, experiment with different rates of fuel consumption (remembering that the initial amount of fuel in the rocket in each case is the same), to discover the effect on the trajectory. Are some fuel consumption rates more efficient than others at reaching a particular altitude?

4.3 **Early engine shutdown:** With the same parameters as in Section 4.3, what happens if the engines shutdown (due to some technical fault) after only 10 s rather than the full 90 s burn?

4.4 **Diffusion from an infinite source:** By altering the boundary conditions in the numerical implementation discussed in Section 4.9 show how diffusion of Al into a GaAs quantum well surrounded by infinitely wide $Ga_{1-x}Al_xAs$ barriers eventually leads to a concentration x of Al in the well.

4.5 **Ion implantation and time dependent diffusion coefficients:** A 100 Å GaAs quantum well surrounded by infinitely wide $Ga_{1-x}Al_xAs$ barriers is implanted with an ionic species which produces sufficient vacancies in the crystal lattice to induce diffusion of Al from the barriers into the GaAs well at a rate described by a constant diffusion coefficient of 1 $\mathring{A}s^{-1}$. The diffusion is allowed to proceed for 10 s after which the semiconductor is subject to a rapid thermal anneal which provides sufficient heat energy for the crystal to repair itself, with the number of vacancies decreasing exponentially with a decay constant of 10 s. Assuming that the diffusion coefficient is proportional to the number of vacancies, show what happens to the profile of the Al distribution.

Projects

4.1 **Rocket trajectory under Newton's law of gravity and the escape velocity:** In the examples in the text the gravitational force on the rocket was taken simply as $m(t)g$. Remove this approximation by including gravity as given by Newton's inverse square law. By numerical experiment deduce the (vertical) speed which must be attained for the rocket to escape the Earth's gravitational field.

4.2 **Multiple-stage launch vehicles:** Show how the model for the trajectory of rockets can be extended to include multiple-stage rockets, i.e. when one rocket has used all its fuel, part of the vehicle is jettisoned and a smaller rocket started. Illustrate with examples using suitable chosen parameters.

4.3 **Simple pendulums and harmonic motion:** Write a computer program to simulate the harmonic motion of a simple pendulum in the Earth's gravitational field. Extend the simulation to include the air resistance on the 'bob' (the mass) at the end of the string. Show how the latter will lead to damped oscillations.

4.4 **Diffusion of dopants in δ-doped semiconductors:** Modern semiconductor devices are often 'modulated doped', which means the amount of either n- or p-type doping depends on the position within the semiconductor. Furthermore, this doping can be introduced into the semiconductor in very thin layers known as 'δ-layers' (after the

Dirac δ-function). One typical dopant for GaAs is Be, given that an initial 10 Å wide δ-layer of doping level 10^{18} cm^{-3} can end up with a full-width at half-maxima of 40 Å by the end of the growth process (which may take 4 hours), deduce the (constant) diffusion coefficient.

4.5 **Electron energy levels in diffused quantum wells:** Consider diffusion of the Al from Ga$_{1-x}$Al$_x$As barriers into a 100 Å GaAs quantum well with a suitably chosen constant diffusion coefficient. Given that the conduction band edge of the Ga$_{1-x}$Al$_x$As alloy is given by:

$$V = 0.67 \times 1587x \quad \text{meV}$$

calculate the corresponding potential profiles. Solve the latter for electrons of effective mass $0.067m_0$ using the shooting method developed in Project 1.1.

5

Stochastic Simulations

5.1 STOCHASTIC OR MONTE CARLO SIMULATIONS

Stochastic simulations are simulations in the sense that they are computational attempts to reproduce natural processes, just as the simulations described in the previous chapter. However, the word 'stochastic' implies that the results of these simulations are non-deterministic—the output from any one simulation is not predetermined, and subsequent executions of identical sourcecode[1] do not produce the same answers. The results of many simulations are combined to produce a statistical representation of the outcomes, for example, the mean (life)time that the electron will remain in that eigenstate is 12 ps.

This class of simulations are also known as 'Monte Carlo' simulations, and yes, this alternative name is derived from the place; more specifically the random processes around which the popular gambling games are based. The author's experience is that the words 'Monte Carlo simulations' are usually associated with electron scattering processes in semiconductors. However, the world of stochastic simulations is very rich and varied, and all the remaining chapters of this book are dedicated to various classes of stochastic simulation—the common link between them being *attempts to simulate natural processes using computational models, with the wide variety of initial and environmental conditions being generated through the use of random numbers.*

---

[1] There is one caveat here to do with the random number generator, see the next section.

5.2 RANDOM NUMBERS

The generation of random numbers by a logical deterministic machine, such as a desktop computer, is more complicated than might seem apparent. However, luckily for the programmers of today, there are standard functions, available in most common programming languages, that can be employed. Such random number generators are not truly random, after initialisation they produce a very long sequence of numbers, but if the same computer program is executed, with the same initialisation, then the same sequence of numbers is generated. The C-code below gives an example of how the random number generator rand() is used:

C Code
for Generating
Random
Numbers

```
#include <math.h>
#include <stdio.h>
#include <stdlib.h>
#define seed 1              /* random number sequence seed */

main()
{
float    r;        /* the random number */
int      i;        /* the loop index */

srand(seed);                 /* initialise random number sequence */

for(i=0;i<1000;i++)
{
 r=(float)rand()/RAND_MAX;                 /* create random number */
 printf("%f\n",r);                         /* print random number */
}

}/* end main */
```

The function srand() performs the initialisation of the random number sequence. It requires an initial starting number called the 'seed' to be passed to it and subsequent calls of the function rand() then return random numbers. In this implementation rand() actually returns an integer between 0 and RAND_MAX (which itself is defined in the header file stdlib.h), hence casting the random integer into a float and dividing by RAND_MAX gives a real number between 0 and 1—which is perhaps the most convenient form for implementing in a range of applications.

The effect of the seed number is illustrated in Table 5.1. The first three *different* values of the seed number produce three different sequences of random numbers. However, when returning to the first value of the seed, the first sequence is repeated. Thus repeated simulations which utilise sequences of random numbers should have different seeds, otherwise the required stochastic nature of the simulations will be negated.

Real hardworking stochastic simulations usually take the seed from some constantly altering variable, such as the time, so that the sequence of random numbers really is different (nearly) every time. This can be implemented by the incorporation of a few additional lines of sourcecode which read the system time (for example) and take, say,

Table 5.1 The first 10 random numbers produced for various values of the seed

seed=1	seed=2	seed=3	seed=1
0.840188	0.700976	0.561380	0.840188
0.394383	0.809676	0.224983	0.394383
0.783099	0.088795	0.393092	0.783099
0.798440	0.121479	0.443938	0.798440
0.911647	0.348307	0.285041	0.911647
0.197551	0.421962	0.144781	0.197551
0.335223	0.699805	0.563555	0.335223
0.768230	0.066384	0.864679	0.768230
0.277775	0.587482	0.895402	0.277775
0.553970	0.642966	0.230805	0.553970

the last two digits as the seed. Such code is machine and operating system specific and, hence, will not be dwelt on any further. It is sufficient for now, with the limited number of simulations required in this work, just to set the seed as a one-digit integer, and increment it each time the simulation is repeated.

5.3 HOW RANDOM IS RANDOM?

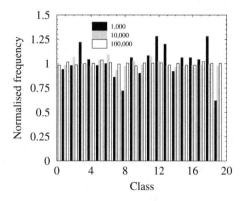

Fig. 5.1 A frequency histogram showing the normalised rate of occurrence of random numbers generated by the function rand() after initialisation with the seed=1. The histogram shows data for sequences of 1,000, 10,000 and 100,000 random numbers with the range divided into 20 classes of equal width

Fig. 5.1 shows the distribution of random numbers across twenty classes of equal width covering the entire range. The frequencies have been normalised by dividing the actual number in the class, by the number that would be expected if they were distributed entirely uniformly. Thus, if there were no bias towards any particular section of the range, then all the 20 chosen classes would have normalised frequencies equal to 1. It can be seen, by studying Fig. 5.1, that although there are quite large discrepancies from this for the first set of 1,000 numbers, as the set increases in size there is a convergence to this optimum distribution.

Thus, it can be concluded that the functions used *here* to generate random numbers are well distributed over the range 0 to 1, and simulations that use them should not *artificially* favour any particular outcome.

5.4   MONTE CARLO SIMULATIONS OF ELECTRON SCATTERING

Consider a group or collection[2] of electrons in a particular eigenstate. Consider also, that the eigenstate is in close proximity to other states of *lower* energy. Then if the system were allowed to respond naturally, the electrons will eventually, one-by-one, scatter (jump) into either of the two lower energy states, see Fig. 5.2.

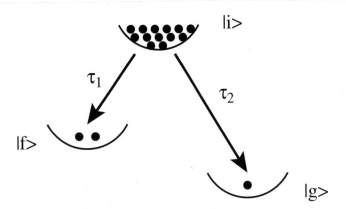

Fig. 5.2   Carrier (electron) scattering channels

The characteristic times τ_1 and τ_2 represent the average time it takes before any one electron scatters from the initial state $|i\rangle$ to the final states $|f\rangle$ and $|g\rangle$ respectively. The reciprocals of these 'lifetimes' give a measure of how many electrons scatter per unit time—*the scattering rates*. Thus if $1/\tau_1$ is twice as large as $1/\tau_2$ then this means that twice as many electrons will scatter into the final state $|f\rangle$ as will into the final state $|g\rangle$.

This system is really quite simple and can be solved analytically: The proportions of carriers that scatter into either of the final states are given by the relative proportions of

---

[2]Often referred to in Monte Carlo language as an 'ensemble'.

the scattering rates, i.e.

$$\text{number of carriers scattering through channel 1} \propto \frac{1/\tau_1}{1/\tau_1 + 1/\tau_2}$$

and

$$\text{number of carriers scattering through channel 2} \propto \frac{1/\tau_2}{1/\tau_1 + 1/\tau_2}$$

Thus, if the characteristic times for scattering through each channel are equal, i.e. $\tau_1 = \tau_2$, then 50% of the electrons will scattering into each of the final states. If the *scattering rate* through channel 1 is 3 times higher than that through channel 2, i.e. $1/\tau_1 = 3/\tau_2$, then 75% of the electrons will scatter through channel 1 and 25% through channel 2.

Hence, after all n_i carriers in the initial state $|i\rangle$, have scattered, the number of carriers in each of the final states $|f\rangle$ and $|g\rangle$ are given by:

$$n_f = n_i \frac{1/\tau_1}{1/\tau_1 + 1/\tau_2} \qquad (5.1)$$

$$n_g = n_i \frac{1/\tau_2}{1/\tau_1 + 1/\tau_2} \qquad (5.2)$$

The *simulation* of this system would involve creating a computational model (writing a computer program) which mimics the physical processes. For example, a number of electrons could be placed in a given state, and in a given time interval Δt, say, the simulation could involve considering each electron in turn and making a decision (based on the result of a random number) as to whether the electron scatters or not, and if it did, which scattering channel it followed. In such a model, the scattering rates are interpreted as *probabilities*. In an *introductory treatise*, the approximations could be made that if $\Delta t \ll \tau$, then[3]:

probability, P_1, scattering through channel 1 in time interval $\Delta t = \dfrac{1}{\tau_1}\Delta t$ (5.3)   *Scattering Probabilities*

and similarly:

probability, P_2, scattering through channel 2 in time interval $\Delta t = \dfrac{1}{\tau_2}\Delta t$ (5.4)

Thus, the simulation could involve making the decision as to whether the electron scattered through channel 1 or 2 or did not scatter at all, by choosing a random number r, say, between 0 and 1 and making a selection based on the value of r, for example:

$$0 < r < P_1 \quad \Rightarrow \quad \text{scattering channel 1} \qquad (5.5)$$
$$P_1 < r < P_1 + P_2 \quad \Rightarrow \quad \text{scattering channel 2} \qquad (5.6)$$
$$P_1 + P_2 < r < 1 \quad \Rightarrow \quad \text{no scattering} \qquad (5.7)$$

Selecting the Scattering Process

---

[3]Note also, that if there are several scattering channels available, the *total* probability should remain less than 1, i.e. $\sum_i P_i < 1$. This can always be ensured in this formalism by reducing Δt. Note further that the results of any simulation should be independent of the value of Δt, which, after all, is just a computational parameter that has been introduced—it has no physical significance—see Fig. 5.6 and the corresponding discussion.

Algorithm 5.1 Algorithmic description of a Monte Carlo model of electron scattering

$e \Leftarrow 1.60219 \times 10^{-19}$ {Define physical constants, for example}

$N \Leftarrow 100$ {Define system parameters, number of electrons}

$T1 \Leftarrow 30\text{ps}$ {Lifetime for scattering into $|1\rangle$}

$T1 \Leftarrow 60\text{ps}$ {Lifetime for scattering into $|2\rangle$}

while there are still some electrons in level 1 **do**

    Increment clock

    for i=1 to N **do**

        Generate random number r

        Decide whether to scatter (and through which channel) or not

        Update the counters n_i, n_f and n_g tracking the electron populations

    end for

    Output final number of electrons in each state

end while

An algorithmic description of this simulation is given in Algorithm 5.1 and a C implementation is given below:

C Code

Monte Carlo Simulation of Electron Scattering

```
#include <math.h>
#include <stdio.h>
#include <stdlib.h>
#define dt 1              /* time interval Delta t, in ps */
#define N 100             /* number of electrons */
#define T1 30             /* channel 1 lifetime, in ps */
#define T2 60             /* channel 2 lifetime, in ps */
#define seed 1            /* random number sequence seed */


main()
{
float    P1,P2;           /* the scattering probabilities */
float    r;               /* a randomly generated number */
float    t=0;             /* the time */
int      ni=N;            /* number electrons in i at t=0 */
int      nf=0;            /* number electrons in f at t=0 */
int      ng=0;            /* number electrons in g at t=0 */
int      i;               /* index over electrons */
```

Calculate Scattering Probabilities

```
P1=(float)dt/T1;P2=(float)dt/T2;              /* calc. probabilities */

srand(seed);              /* initialise random number sequence */
```

Loop over Time

```
do
{
  t+=dt;
  for(i=0;i<ni;i++)
  {
    r=(float)rand()/RAND_MAX;        /* generate random number */
    if(r<P1){ni--;nf++;}             /* scattering channel 1? */
    if((r>P1)&&(r<(P1+P2))){ni--;ng++;}     /* channel 2? */
  }
  printf("%f %i %i %i\n",t,ni,nf,ng);        /* output populations */
}while(ni>0);

}/* end main */
```

Continue until all Electrons have left Initial State

5.5 TESTS AND LIMITS

Perhaps the simplest test is just to consider a 2-level system—the scattering from $|i\rangle$ directly to $|f\rangle$, without the additional complication of the second final state $|g\rangle$. This can be achieved quite simply by making the lifetime for scattering through channel 2 very long (10000 ps, for example). Fig. 5.3 shows the results of this first test calculation.

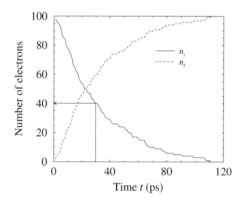

Fig. 5.3 The number n_i and n_f of electrons in the initial $|i\rangle$ and final $|f\rangle$ states as a function of time for $\tau_1 = 30$ ps and τ_2 effectively infinite

The obvious and expected result, that the number n_i of electrons in the initial state decreases and the number n_f of electrons in the final state increases, is clear from the figure. The randomness of the simulation is illustrated by Fig. 5.4 which shows several simulations identical to the previous figure, but with different values for the random number seed.

The physical meaning of the lifetime (in this case τ_1) can also be obtained from Fig. 5.3. When the time passed equals the lifetime, i.e. when $t = \tau_1$, the number n_i of electrons in the initial state has fallen to approximately e^{-1} of its initial value, as can be seen by

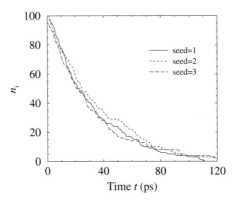

Fig. 5.4   The number of electrons n_i in the initial state $|i\rangle$ as a function of the time, for $\tau_1 = 30$ ps and for three different values of the random number seed

the interpolating arrow in the figure. Indeed, in a simple two-level system as in these first simulations, the population of a level does decay exponentially:

Exponential Decay

$$n_i(t) = n_i(0) \exp\left(-\frac{t}{\tau_1}\right) \tag{5.8}$$

This is reinforced by the data in Fig. 5.5(a) which shows the decaying population of the initial state for three different lifetimes τ_1. Fig. 5.5(b) displays the same data but on log–linear axes—the fact that the data can be fitted very well with straight lines proves the exponential decay nature.

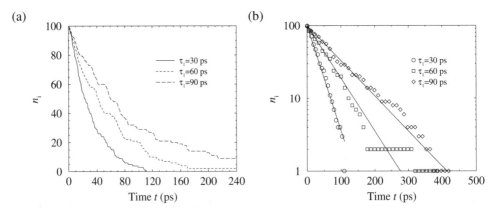

Fig. 5.5   (a) The number of electrons n_i in the initial state $|i\rangle$ as a function of the time, for $\tau_1 = 30$, 60 and 90 ps, (b) the same data on a log scale with corresponding straight line fits of the form $n_i = n_i(0) \exp\left(-t/\tau_1\right)$

Perhaps the final point to confirm in this implementation of a Monte Carlo model is the choice of the time interval Δt. When Δt was introduced, in Section 5.4, the probabilities of scattering through the different channels were derived, in equations (5.3) and (5.4), based on the assumption that the Δt was much less than the characteristic times τ_1 and τ_2.

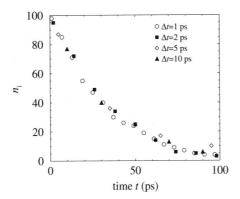

Fig. 5.6 The number of electrons n_i in the initial state $|i\rangle$ as a function of the time t for several different time intervals Δt (for $\tau_1 = 30$ ps)

Fig. 5.6 shows the results of a series of calculations aimed at validating this assumption. The Monte Carlo simulations are similar to those already presented in this section, with τ_2 made very long in order to restrict the system to a single scattering channel and the lifetime τ_1 was chosen, as in some of the previous examples, to be 30 ps. It can be seen that varying the value of Δt (known as dt in the program) from the value used so far (1 ps) to 2, 5 and even 10 ps does not significantly alter the results of the calculations. Thus the approximation made earlier *is* valid and the output from the simulations is independent (within these limits) of the choice of this computational parameter. Further calculations with $\Delta t = 20$ ps did begin to show some deviation from the data in Fig. 5.6.

5.6 SIMULATIONS OF THREE-LEVEL SYSTEMS

Fig. 5.7(a) shows the results of simulations of the full three-level system originally introduced in Section 5.4. In this simulation the characteristic times τ_1 and τ_2 were chosen to be equal and set at 30 ps. As would be expected intuitively, the time dependence of the numbers of electrons in the final states $|f\rangle$ and $|g\rangle$ are equal (to within statistical fluctuations).

Fig. 5.7b shows the effect of one scattering channel being preferred to the other—simulated by choosing $\tau_1 = 30$ ps and $\tau_2 = 60$ ps. In this particular simulation of 100 electrons, the final numbers in states $|f\rangle$ and $|g\rangle$ are 60 and 40 respectively, which differ somewhat from what might be expected from the analytical predictions earlier. Recall equations (5.1) and (5.2) and setting $\tau_1 = 30$ and $\tau_2 = 60$ ps, as in the simulations here then:

$$\frac{n_f}{n_i} = \frac{1/30}{1/30 + 1/60} = \frac{2}{3}$$

(5.9) *Analytical Solutions*

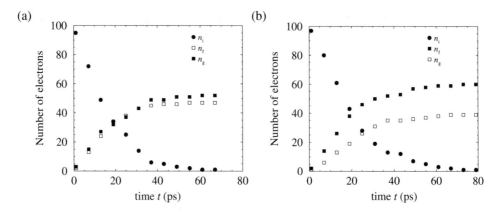

Fig. 5.7   The number of electrons in the two final states $|f\rangle$ and $|g\rangle$ as a function of the time, for (a) $\tau_1 = \tau_2 = 30$ ps, and (b) $\tau_1 = 30$ ps and $\tau_2 = 60$ ps

$$\frac{n_g}{n_i} = \frac{1/60}{1/30 + 1/60} = \frac{1}{3} \tag{5.10}$$

The difference does arise from statistical fluctuations. Hitherto, the simulations have been based on a set of just 100 electrons, and perhaps this is only large enough to give 'ball park' predictions. Fig. 5.8 shows the results of simulations with larger numbers of electrons which are otherwise identical to that in Fig. 5.7(b). The graph shows just the time dependence of the number of electrons in the initial state $|i\rangle$, *normalised* to the total number of electrons. It can be seen that there is some 'roughness' in the data due to statistical fluctuations for the original set of $N = 100$ electrons. However, the larger data sets of 1,000 and 10,000 both show a smooth exponential decay—the statistical fluctuations are smaller as a proportion.

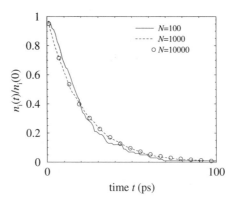

Fig. 5.8   The number of electrons n_i in the initial state $|i\rangle$ as a function of the time t for simulations of different numbers N of electrons

Table 5.2 summarises the proportions of electrons in the final states $|f\rangle$ and $|g\rangle$ from this same series of calculations. It is clear that the predictions of the simulations do converge to the analytical (exact) solutions of 2/3 (0.666) and 1/3 (0.333), thus further validating this simple implementation of a stochastic simulation.

Table 5.2 The effect of increasing the number of electrons on the final state populations n_f and n_g. Notice, that the increasing size of the set allows an additional decimal place to be added to the final results of the simulations

N	n_f/N	n_g/N
100	0.60	0.40
1,000	0.656	0.344
10,000	0.6610	0.3390
100,000	0.66606	0.33394
1,000,000	0.666375	0.333625

Numerical Estimates

Thus a method has been derived for simulating the scattering of charge carriers between multiple energy levels. Although these introductory examples have been restricted to one initial and two final levels, the technique can be readily extended to many more levels, with several sequential scattering channels—see the Tasks and Projects at the end of this chapter.

5.7 CHANGING PROBABILITIES AND PAULI EXCLUSION

Hitherto, the characteristic lifetimes (for example τ_1 and τ_2 in the previous sections) describing the rate at which electrons have entered scattering channels, have been taken as constants. There is no restriction within the Monte Carlo method to require this, and the lifetimes τ can be functions of virtually anything.

One important example of such a phenomenon does indeed occur in systems of electrons (or holes) scattering between different energy levels. The Pauli Exclusion Principle limits the number of electrons in any one state to two (ignoring electron spin). Now usually an energy level within a solid state system is actually composed of many states close together to form a band. However, the number of states is indeed finite and band filling is a commonly observed phenomena.

Such filling of the final state would have the effect of *reducing* the number of states available for electrons to scatter into, and in turn this could prevent a scattering event from happening. This would manifest itself in Monte Carlo simulations, of the type described here, as a continuous reduction in the scattering rate (an increase in the characteristic lifetime).

Fig. 5.9 shows the results of calculations to illustrate this effect. The lifetime τ_2 for scattering through channel 2 was again made very large in order to naturally limit the system to two levels, i.e. one initial $|i\rangle$ and one final $|f\rangle$ state. One thousand electrons

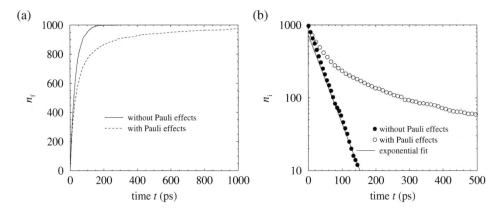

Fig. 5.9   (a) The number of electrons n_f in the final state $|f\rangle$ as a function of the time, with and without the changing scattering probability due to Pauli's exclusion principle and (b) the number n_i in the initial state $|i\rangle$, from the same simulations, on log-linear axes. The straight line fit to the data 'without Pauli' effects indicates exponential decay

were employed in the simulation and it was assumed that the final level had only 1000 states (spaces for electrons). The probability P_1 for scattering into the final state was continually adjusted according to the number of electrons in that state. For the sake of an example, the following choice was made:

Accounting for Pauli Exclusion

$$P_1 = \left(1 - \frac{n_f}{N}\right)\frac{\Delta t}{\tau_1} \tag{5.11}$$

Thus, initially when there were no electrons in the final state, $n_f = 0$ and $P_1 = \Delta t/\tau_1$ as before. This is reflected in the data of Figs. 5.9(a) and (b) where the early changes in the electron populations are similar in the simulations with and without the Pauli effect. However, as time passes and the number of electrons in the final state $|f\rangle$ increases, the probability P_1 that an electron will scatter into that state *decreases*. Thus, the rates at which the electrons enter the final state and leave the initial state both decrease as the time increases, see Figs. 5.9(a) and (b) respectively. In summary, accounting for Pauli exclusion, in this example, reduces the rate at which electrons transfer between the states.

The solid symbols in Fig. 5.9(b) show data, produced as before, from a simulation with a constant scattering probability. And as before, a straight line fit to this data (the solid line) indicates that the number of electrons in the initial state decreases *exponentially*. The data plotted with the open symbols was obtained with the varying scattering probability, as described by equation (5.11). Perhaps the most important conclusion of this section is that the latter data cannot be fitted with a straight line, thus the decay of the electrons in the initial state is *non-exponential*.

Although this example of non-exponential decay in a two-level system could probably be solved analytically, it has served to illustrate how quite complicated effects can be incorporated, rather simply, into a Monte Carlo simulation. Monte Carlo simulations, of carrier dynamics, and other related problems, are therefore used quite extensively, and offer an opportunity to investigate and explore time-dependent aspects of large collections of interacting particles.

Summary

It has been shown that the scattering of electrons between different energy levels can be simulated, as a function of time, by selecting scattering channels on the basis of random numbers, between 0 and 1. If a characteristic time τ is assigned to each scattering channel, then the probability P that any one electron will scatter in a time interval Δt can be approximated by:

$$P = \frac{\Delta t}{\tau}$$

Example simulations showed this to be valid provided $\Delta t \ll \tau$. The scattering channel would be selected if the random number $r < P$, for example. Further simulations demonstrated that if τ described the rate at which electrons scattered out of a particular level and τ was a constant, then the number of electrons at any time t remaining in the level could be written as:

$$n(t) = n(0) \exp\left(-\frac{t}{\tau}\right)$$

i.e. constant characteristic times describe exponential decay processes. Discussions of Pauli's exclusion principle and the limitation of a finite number of final states led to a simulation being developed in which the scattering probability varied with time. It was shown that such effects lead to non-exponential dependencies.

Tasks

5.1 **The randomness of a random number generator:** Using the computer code given in Section 5.2 (adapt if necessary) generate sets of 1,000, 10,000 and 100,000 random numbers and plot frequency histograms using 20 classes between 0 and 1 (as in Section 5.3). Discuss the randomness of the random number generator on *your* computer.

5.2 **Comparison of simulations with analytical solutions:** Adapt the simulation developed in Section 5.4 to represent a two-level system, with all the carriers initially in one of the levels. Choose an appropriate time interval Δt and perform simulations for values of the characteristic lifetime τ from 1 to 1000 ps. Calculate the characteristic lifetime produced by the simulation (by measuring the slope of the $\ln n$ versus time t curve). How many electrons need to be simulated, in each case, in order for the simulation to reproduce the original (input) lifetime τ to within 0.1%?

5.3 **Further comparison of simulations with analytical solutions:** Use the simulation developed in Section 5.4 to represent a three-level system, with all the carriers initially in one of the levels. Choose an appropriate time interval Δt and perform simulations with the characteristic lifetime for one scattering channel fixed at 30 ps and the second lifetime taking several values between 3 and 300 ps. Deduce the ratio of carriers distributed between the two final states. How many electrons need to be included within the simulation, in each case, in order to reproduce the analytical solution for the branching ratios to within 0.1%?

5.4 **Three-level sequential scattering:** Adapt the simulation developed in Section 5.4 to represent *sequential* scattering in a three-level system, with all the carriers in just one of the levels initially. The sequential limitation implies that the only possible scattering mechanisms are $|1\rangle \rightarrow |2\rangle$ and $|2\rangle \rightarrow |3\rangle$. Show how the number of electrons in levels $|2\rangle$ and $|3\rangle$ vary with time for a fixed characteristic time $\tau_{1 \rightarrow 2}$ of 30 ps and various times $\tau_{2 \rightarrow 3}$ from 3 to 300 ps.

5.5 **Three-level branching and sequential scattering:** Repeat the range of simulations as in the previous task, but with the additional scattering channel of $|1\rangle \rightarrow |3\rangle$. By deducing the number of electrons that take either of the two possible routes from the initial state $|1\rangle$ to the final state $|3\rangle$, show how, for short characteristic lifetimes $\tau_{1 \rightarrow 2}$ and $\tau_{2 \rightarrow 3}$, the two-stage sequential scattering can compete with a slower single stage scattering from $|1\rangle$ to $|3\rangle$.

Projects

5.1 **Random number generator:** Write a mathematical function which could be used (with a seed) to generate random numbers with 4 figure accuracy, between 0 and 1. Plot a frequency histogram, as in Task 5.1, for sets of 1,000, 10,000 and 100,000 numbers and discuss the 'randomness' of the sequence.

5.2 **Monte Carlo simulation of traffic flow:** Create a simulation of one-dimensional (and uni-directional) traffic flow along 10 km of straight road (without any junctions). Generate cars at random time intervals with random speeds between say 60 and 120 km/h. Assume that the driver of each car will adjust their speed (up to the maximum of 120 km/h) according to the distance between themselves and the car in front. In particular, assume that changes in speed are governed by the well known equation of motion:

$$v^2 = u^2 + 2as$$

where s is the distance covered by a car accelerating at a rate a from an initial speed u to a final speed v. Starting from a very low rate of cars using the road, say 1 car per minute, increase the rate at which cars enter and use the simulation to describe what happens. State any other assumptions that are needed to get the simulation working.

5.3 **Numerical integration with a Monte Carlo simulation:** Show how, with a specific example, a definite integral could be evaluated with a method involving random numbers. Compare the result with a known analytical solution.

5.4 **Other simulations:** Create a stochastic simulation of a real world system. Illustrate its behaviour with a range of examples.

6

Percolation Theory

6.1 COMPLEX MANY-BODY INTERACTING SYSTEMS

Many things in the world and the universe around us consist of enormous numbers of interacting particles and appear to be extremely complex. For example, the stars that make up galaxies, populations of plants and animals, and right down to microscopic length scales and the uncountable numbers of atoms which make up condensed matter systems. In some cases the behaviour of the collective can be predicted, interpreted or modelled, by considering the interactions between the individual entities. This is certainly true of models of galaxy evolution. Each star interacts with all the other stars through the single long range force of gravity, and the shape of a galaxy can be deduced by calculating the time dependence of the positions of all the stars. It's a long calculation but it can be (and is being) done.

A colony of ants is a complex many-body interacting system, with many more types of interaction between the individuals than a collection of stars which make up a galaxy. Thus, an ant colony is more complicated and difficult to model than a galaxy.

'Percolation theory' is an approach to modelling complex many-body interacting systems, like the ant colony but not the galaxy, by simplifying the interactions between the individuals, and allowing the system to evolve (find its own equilibrium) within a stochastic simulation. The collection of individuals is usually represented in the simplest form imaginable, such as a two-dimensional grid, and the interactions could be limited to nearest neighbours only, and be simplified to the extent that either there is an interaction or there is not an interaction.

Percolation theory[1] can be applied to any system which can be reduced to very simple interactions between the individuals. It is applied to situations as diverse as forest fires, oil extraction, the evolution of species, colonisation of continents, magnetism and, of course, filter coffee. Percolation theory is best explained using an example.

6.2 DISEASE PROPAGATION

A human population within some kind of closed system, whether it be because of physical boundaries, such as an ocean, or political boundaries, or even the entire planet itself, is a complex interacting many body system. And, although the development[2] of human culture and socio-economic systems are governed by a myriad of factors and lie outside the realms of an introductory mathematical modelling text, some insight into some of the features of human society may be gained from the application of percolation theory.

One example might be in the important area of the propagation of a disease, either a virus or a bacterium, in a partially immunised population. To a bacterium a person is just a 'host', a place to be or a place to live and from this viewpoint a human population is a much simpler place. There are fertile islands (unimmunised people) and their are barren islands (immunised people), and the sole aim in life of a bacterium is to colonise as many islands as possible.

A bacterium achieves its aim by moving sequentially from one host to another. In the main they will transfer from person to person who are in close proximity to one another. In terms of percolation theory a population could be represented by a two-dimensional array of people, which initially is completely disease free, see the left-hand diagram in Fig. 6.1, and consists of a proportion x who are immunised against the disease and hence a proportion $(1 - x)$ who are not, see the right-hand diagram in Fig. 6.1.

When one person becomes infected, either from an air-borne spore, or travelling to an area where the bacterium is present, the cause really doesn't matter. This can be represented by an individual being chosen at random and given the infection, see Fig. 6.2. As a starting point, the model could be constructed to say that the bacterium is then able to infect all adjoining hosts who are not immunised. This is equivalent to saying that, in this model, each person typically comes into close enough proximity to pass the disease to eight other people. This assumption simplifies the model for now and will allow a working implementation to be developed, although it is not a limitation and it will be shown later how this 'coordination number' can be changed to represent other degrees of mixing between the people in a population.

It can be seen that if the disease is passed by close contact, then in the first example, Fig. 6.2(a), the disease will be able to propagate through nearly the entire unimmunised (white boxes) population. However, if the source of the disease only comes into contact with a few unimmunised people who themselves only mix with immunised non-carrier

---

[1]If you want to know much more than is presented here, see D. Stauffer and A. Aharony, 'Introduction to Percolation Theory', Taylor and Francis, London, Second Edition, 1994.
[2]In a scientific context 'development' might be taken as meaning 'time-dependence'.

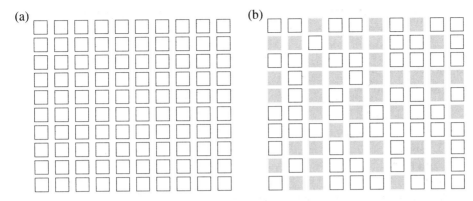

Fig. 6.1 (a) A representation of a human population, and (b) before the disease arrives; the uninfected (white boxes) and the immunised (grey boxes) for an immunisation rate of $x = 0.5$ (50%)

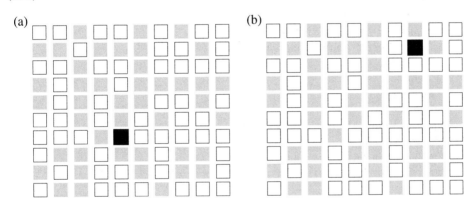

Fig. 6.2 Two examples of the onset of infection of a population from a single individual

(grey boxes) individuals, as in the second example, Fig. 6.2(b), then the disease will not be widespread.

6.3 COMPUTATIONAL IMPLEMENTATION

The particular application of percolation theory discussed in the previous section could be described by Algorithm 6.1.

Algorithm 6.1 Algorithmic description of a percolation model of disease propagation

---

$N \Leftarrow 10$

$\{\text{Population}=N^2\}$

$\{\text{Immunise a randomly chosen fraction of the population}\}$

for i=1 to N **do**
  for j=1 to N **do**
    Choose a random number r
    if $r < x$ **then**
      immunise person $P[i][j]$
    end if
  end for
end for
Select a person at random and infect them

$\{\text{Now let the disease propagate}\}$

while number of infected people increases **do**
  for i=1 to N **do**
    for j=1 to N **do**
      if $P[i][j]$ is infected **then**
        infect all unimmunised neighbours
      end if
    end for
  end for
end while
Output coordinates of immunised people
Output coordinates of uninfected people
Output coordinates of infected people

A C-code implementation of this algorithm could be:

C Code
Percolation
Theory
Model of
Disease
Propagation

```
#include <math.h>
#include <stdio.h>
#include <stdlib.h>
#define N 10          /* population is N*N */
#define x 0.5         /* proportion immunised */
#define seed 1        /* random number sequence seed */
```

```
main()
{
int      min();            /* ensures i and j >0 */
int      max();            /* ensures i and j <N-1 */
double   r;                /* the random number */
int      P[N][N];          /* the population */
int      i;                /* index over population */
int      j;                /* index over population */
int      n=0;              /* number of infected people */
int      n_last=-1;        /* previous (last) value of n */
int      t=0;              /* iteration counter */
char     file[16];         /* string for storing filename */
FILE     *Fimmunised;      /* Immunised population file */
FILE     *Funinfected;     /* Uninfected population file */
FILE     *Finfected;       /* Infected population file */



srand(seed);               /* initialise random number sequence */



for(i=0;i<N;i++)           /* immunise fraction 'x' of population */
  for(j=0;j<N;j++)
  {
  r=(float)rand()/RAND_MAX;                  /* random number */
  if(r<x)P[i][j]=0;        /* immunise */
  else P[i][j]=1;          /* leave uninfected */
  }



P[3][4]=2;                 /* infect the first individual manually */
```

Immunise Fraction of Population

Source of Disease

```
while(n>n_last)            /* keep going until as many people */
{                          /* infected as possible */
 n_last=n;n=0;             /* initialise counter */
 t++;                      /* counter tracking number of iterations*/
 for(i=0;i<N;i++)                    /* consider all people */
  for(j=0;j<N;j++)
  {
    if(P[i][j]==2)                   /* if infected */
    {
    n++;                             /* count how many infected */
    /* now infect neighbours */
    if(P[min(i-1)][min(j-1)]==1)P[min(i-1)][min(j-1)]=2;
    if(P[min(i-1)][j]==1)        P[min(i-1)][j]=2;
    if(P[min(i-1)][max(j+1)]==1)P[min(i-1)][max(j+1)]=2;
    if(P[i][min(j-1)]==1)        P[i][min(j-1)]=2;
    if(P[i][max(j+1)]==1)        P[i][max(j+1)]=2;
    if(P[max(i+1)][min(j-1)]==1)P[max(i+1)][min(j-1)]=2;
    if(P[max(i+1)][j]==1)        P[max(i+1)][j]=2;
    if(P[max(i+1)][max(j+1)]==1)P[max(i+1)][max(j+1)]=2;
    }
  }

/* open external data files for writing data */


sprintf(file,"immunised%i.r",t);Fimmunised=fopen(file,"w");
sprintf(file,"uninfected%i.r",t);Funinfected=fopen(file,"w");
sprintf(file,"infected%i.r",t);Finfected=fopen(file,"w");

for(i=0;i<N;i++)
 for(j=0;j<N;j++)
 {
  if(P[i][j]==0)fprintf(Fimmunised,"%i %i\n",i,j);
  if(P[i][j]==1)fprintf(Funinfected,"%i %i\n",i,j);
  if(P[i][j]==2)fprintf(Finfected,"%i %i\n",i,j);
 }

fclose(Fimmunised);
fclose(Funinfected);
fclose(Finfected);
}  /* end while */

}/* end main */

int min(i)        /* returns value of integer limited to 0 (zero) */
```

```
int      i;
{
 if(i<0)i=0;
 return(i);
}

int max(i)            /* returns value of integer limited to N-1 */

int      i;
{
 if(i>(N-1))i=N-1;
 return(i);
}
```

There are some features worth commenting on in this implementation. In particular, the individuals within the population are represented by an array and the value stored at that array address describes the diseased state of the person. For example, if the value is '0' the person is said to be immunised and therefore unable to be infected, if the value is '1' the person is uninfected and if the value is '2' then the person is infected and carrying the disease.

After infecting an individual (manually, by a simple choice), the program considers each individual in turn, and if infected (if(P[i][j]==2), it increments the counter (n++) tracking the number of people infected, and then infects all unimmunised (P[i][j]==1) neighbouring people. The program repeats this, within a while loop, until the number of infected people on the current pass, n, is the same as that on the previous (last) pass, n_last. In this form, the C implementation prints out the coordinates of all the people in files according to their state.

Fig. 6.3 shows the results of allowing the disease to propagate through the population from the starting point in Fig. 6.2(a), using this C implementation. Fig. 6.3(a) shows the population after just one pass, which could be interpreted as the time interval for each person to have contact with their 8 'close' friends, relatives and colleagues—perhaps one day. It can be seen that, with this degree of mixing, the disease has spread through the entire population in just 4 days. As predicted earlier, the few individuals who are separated from the majority of the population by a network of immunised people, remain uninfected.

6.4   THE DEPENDENCE OF INFECTION UPON IMMUNISATION

As with any stochastic simulation it is important to run it many times in order that the predictions have some statistical basis. To achieve this within this algorithm, the first infected individual would be chosen at random and an additional loop could be included to repeat the simulation many times, with the results just expressed in terms of the number of people infected. Thus the complex interactions within the population are *summarised* with just one number. This is perhaps a fairer reflection of the amount of information that can be extracted from a percolation approach. The individuals are just boxes which have a state which is either 0 (immunised), 1 (uninfected), or 2 (infected). No significance

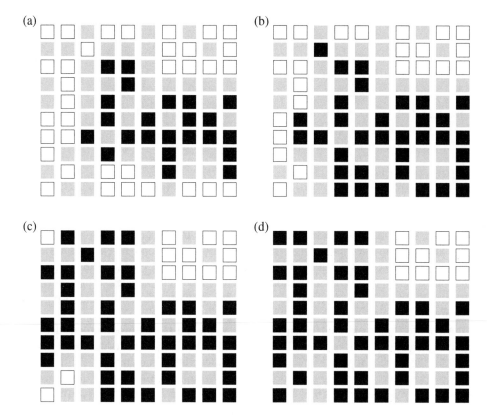

Fig. 6.3   The propagation (a)–(d) of a disease, for the population and source of infection as in Fig. 6.2(a)

can be attached to *which* individuals are infected, but perhaps by averaging the number of infections over a very large population (produced by repeating the simulation for a small population many times), there is some significance to the infection rate—the proportion of unimmunised people who become infected.

One point of interest with regard to a population and the containment of a disease might be the dependence of the number of people infected on the proportion of people immunised (x). Fig. 6.4 shows the results of developing the simulation from the last section to include many simulations for many different immunisation rates, as discussed above[3]. Fig. 6.4(a) shows the number infected in every population of 100 (10×10) people, averaged over 10 and 1000 simulations. The statistical fluctuations in the first of these are clearly evident, although they have virtually disappeared by the time the simulation is repeated 1000 times. Fig. 6.4(b) just confirms that the data are independent of the choice of the value of the random number seed, showing identical results over the whole range of immunisation rates x.

---

[3]The development of this simulation is the subject of Project 6.3.

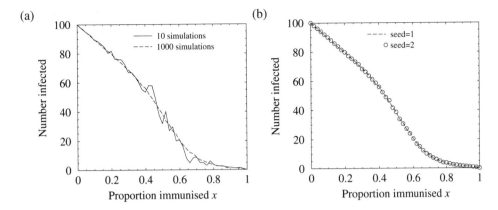

Fig. 6.4   The number of people infected versus the proportion x of the population immunised, for (a) 10 and 1000 simulations for each x, and for (b) 1000 simulations, with two different seeds for the random number sequence

The data itself show that for small x there is a linear dependence of the number of people infected versus the immunisation rate. However, in the second half of the x domain, there is a deviation from linearity and the number of people infected (and its derivative) tends to zero.

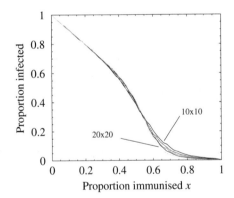

Fig. 6.5   The proportion of people infected versus the proportion x of the population immunised, for populations of 10×10, 12×12, 14×14, 16×16 and 20×20 individuals, all averaged over 1000 simulations

Within these finite populations, although the assumption is made that each individual is coming into close enough proximity to pass the disease to 8 other individuals (given by the natural coordination of a two-dimensional square pattern), in fact, the individuals along the edges of the population have only 5 neighbours and those at the corner only 3. To see if this influences the results in any way, Fig. 6.5 illustrates the results of calculations of the number of people infected for several different sizes of population. It can be seen that there is a small but noticeable effect when x is between 0.6 (60%) and 0.8 (80%). However, the functional form of the number of people infected is the same.

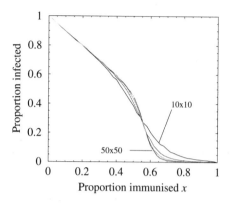

Fig. 6.6   The proportion of people infected versus the proportion x of the population immunised, for populations of 10×10, 20×20, 30×30, 40×40 and 50×50 individuals, all averaged over 1000 simulations

Fig. 6.6 extends the data set from the previous figure to show the results of (rather long) simulations of 30×30, 40×40 and 50×50 populations in comparison with the largest data set from Fig. 6.6. It can be seen that, as the population increases, the curves seem to be converging towards the line:

$$y = 1 - x, \quad \text{for} \quad 0 < x < 0.5 \tag{6.1}$$

For immunisation rates x greater than about 0.5 the proportion of people infected just seems to keep decreasing as the population increases. The behaviour of the curves at these two extremes can be seen more clearly by looking at the data in another way, as in the next section.

6.5   THE LIMIT OF LARGE POPULATIONS AND CRITICAL DENSITIES

So far, the data from these simulations has been viewed in terms of the proportion of the population infected. However, there is another way in which to quantify the data, which perhaps reveals more about the propagation of the disease.

Fig. 6.7 shows the same data as plotted in Fig. 6.5, however, this time the results are displayed in terms of the 'infection rate' defined as:

$$\text{infection rate} = \frac{\text{number of people infected}}{\text{number of unimmunised people}} \tag{6.2}$$

which is equivalent to:

$$\text{infection rate} = \frac{\text{proportion of people infected}}{1 - x} \tag{6.3}$$

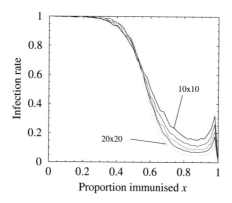

Fig. 6.7 The proportion of *unimmunised* people infected versus the proportion x of the population immunised, for populations of 10×10, 12×12, 14×14, 16×16 and 20×20 individuals, all averaged over 1000 simulations

It can now be seen that, when the immunisation rate *is* less than about 0.4, the infection rate is equal to 1. This means that if less than 40% of the population are immunised against the disease, then the introduction of a single[4] source can lead to the entire unimmunised population being infected. This happens because there are not enough immunised people to form 'walls' to prevent the spread of the disease. However, when the immunisation rate x increases above about 0.6 (60%), there are real benefits. There are now enough immunised people to block the path of the disease and consequently, the simulations show that the infection rate is kept down to about 0.2, i.e. only about 20% of the unimmunised population catch the disease.

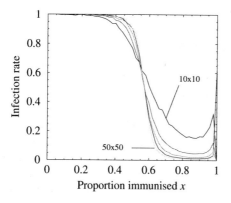

Fig. 6.8 The proportion of *unimmunised* people infected versus the proportion x of the population immunised, for populations of 10×10, 20×20, 30×30, 40×40 and 50×50 individuals, all averaged over 1000 simulations

---

[4]The consequences of multiple sources are the subject of Task. 6.2.

Fig. 6.8 again plots the same data as previously (Fig. 6.6), and again, now in terms of the infection rate. The larger populations show more dramatic behaviour of the infection rate, with the rate converging even more strongly to 1 for immunisation rates x up to about 0.5 and down towards zero for x greater than about 0.6 . The original infection of the population seems to be quickly contained by the large fraction of immunised people who 'block' the transmission of the bacterium. As the population grows in size, the *proportion* that this represents becomes smaller, and hence the behaviour of the curves above $x = 0.6$ in Fig. 6.8.

It can be imagined that in the limit of very large populations there will be a very sharp change from an infection rate of 1 to an infection rate of 0. Extrapolation of the data in Fig. 6.8 shows that this will occur around $x = 0.56$. This point represents a 'phase change' whereupon a population makes a transition from all the unimmunised people being infected, to remaining uninfected. In percolation theory models, this proportion of immunised people is known as a 'critical density'.

6.6 MIXED MAGNETIC PHASES

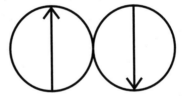

Fig. 6.9   The antiferromagnetic aligning of neighbouring Mn ions

There are a group of semiconductors, in which some of the group II cations are replaced by Manganese (Mn) to form a 'dilute magnetic semiconductor', $Cd_{1-x}Mn_xTe$ is one example. The magnetism of this material is quite complex, in that it consists of mixed paramagnetic and anti-ferromagnetic phases. These mixed phases arise because, although the Mn ions like to align their spins anti-parallel with each other to form an antiferromagnet, see Fig. 6.9, when the concentration x of Mn is low, the ions are often isolated and cannot pair, thus producing regions of paramagnetism, see Fig. 6.10.

When the concentration x of Mn ions is high (x approaches 1—pure MnTe), the Mn ions will certainly be adjacent to several other Mn ions, and hence will readily be able to align antiferromagnetically. However, if the concentration of Mn ions was reduced, a point would arise where some Mn ions would be isolated (have no magnetic neighbours) and so would not be able to pair their spins[5]. These Mn ions would then be able to align their spins in the direction of any external magnetic field, thus contributing to the paramagnetism.

---

[5] Assuming that the interaction only extends to nearest neighbours.

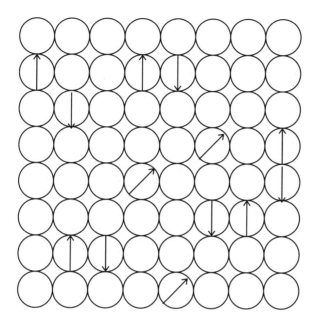

Fig. 6.10 A representation of a dilute magnetic semiconductor, such as $Cd_{1-x}Mn_xTe$, with regions of antiferromagnetic spin-pairing, and regions of free Mn spins exhibiting paramagnetism

A question that might interest the numerical modeller might be: 'How does the concentration of isolated (free) spins vary with the concentration x, of Mn ions?'. The modelling of these materials is ideal for percolation theory and, on the face of it, it is a complex interacting many-body system. However, the interactions can be simplified to: Consider a Mn ion, if it has a neighbouring Mn ion (which is unpaired), then pair the spins of both of them, otherwise leave it alone[6]. And, as with any percolation theory approach, the consequences of these assumptions would be calculated and the results would be compared with experiment—if the agreement is good then its good evidence that the original assumptions are sound too.

The procedure to implement a percolation model of this system would be first to randomly generate a crystal lattice (a population of ions), and make a proportion x of them magnetic, see Fig. 6.11. The latter could be achieved (as in the immunisation procedure of the disease propagation example) by choosing a random number r between 0 and 1, for each ion in turn, and if less than x, give it (the individual) a spin.

The model then proceeds by considering each ion in turn, by following some predefined path, as in Fig. 6.12 and, if that ion is magnetic, pairing it with any unpaired neighbouring magnetic ion. The results of such a procedure, for the initial distribution of magnetic ions in Fig. 6.12, are shown in Fig. 6.13.

---

[6]I am using the principle of simplicity at this point to avoid having to consider three or more magnetic ion spins being locked together.

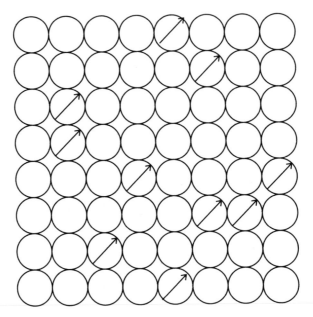

Fig. 6.11 Generating the random positions of the magnetic ions within a host crystal lattice. The concentration x of the magnetic ions was taken as 0.16, which implies 1 in every 6 ions is magnetic

It can be seen that from the original 10 magnetic ions, 6 of them form pairs leaving 4 unpaired. Thus the effective concentration x' of (unpaired) magnetic ion spins is $4/64 \approx 1/16 = 0.0625$ or 6%. As with any stochastic simulation, such a procedure would have to be repeated many times in order to obtain statistically significant results, and then it would have to be repeated for many values of x (the concentration of magnetic ions) in order to achieve the earlier objective.

Simple though this model seems, and remember the assumptions that: (i) two neighbouring unpaired magnetic spins pair, and (ii) only nearest neighbours interact, are enough to explain/predict the paramagnetic behaviour of this class of materials[7] (when performed on the real face-centred cubic crystal lattice with a coordination number of 12, and not the 8 here). The computational implementation of this application form the basis of some of the tasks and projects at the end of this chapter.

---

[7] I've got this far and I'm determined not to cite a scientific paper, but if you really need to know more, look up works to which the author has contributed.

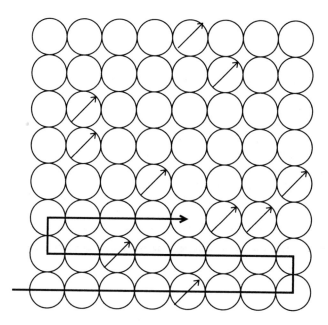

Fig. 6.12   A path through the crystal in which each magnetic ion is considered in turn

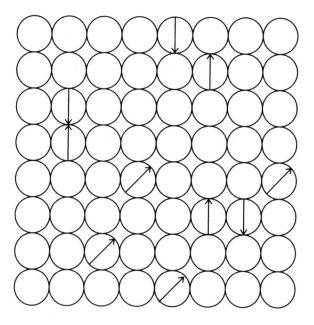

Fig. 6.13   The result of pairing the magnetic spins with unpaired neighbours

Summary

Percolation theory is an approach for simulating the behaviour of complex many-body systems. It is based around reducing the interactions between the particles to their lowest practical form and developing a computational simulation which allows the individuals to interact with each other. The behaviour of the macroscopic system evolves from the behaviour of the individuals.

It was found in the example of disease propagation in a population of living organisms, that there was a critical density for immunisation. Below this point the entire unimmunised population would be infected by the disease, and above this point the vast majority of the population would remain uninfected.

It was also shown how percolation theory could be applied to deduce the relative proportions of the mixed magnetic phases often seen in dilute magnetic solids.

Tasks

6.1 **Population dependence of computational time:** Using the simulation developed in Section 6.3, plot the execution (run) time for simulations of increasingly large $N \times N$ populations, for a variety of immunisation rates x. Is it quicker to run many small simulations (for example, one hundred 10×10), or fewer large (for example, one 100×100) simulations?—*ensure the total number of individuals considered is the same*. Are the infection rates, i.e. the proportion of people infected, the same?

6.2 **Multiple sources of infection:** Adapt the implementation in Section 6.3 to include the possibility of more than one person being infected with the disease at the start. By performing appropriate simulations, show how the infection rate is affected by the inclusion of multiple sources.

6.3 **Development of the simulation of diluted magnetic semiconductors:** Implement the model introduced in Section 6.6 for the mixed phases of diluted magnetic semiconductors. Deduce the effective concentration x' (the concentration of free spins), as a function of the concentration of magnetic ions x.

Projects

6.1 **Coordination number (degree of mixing) of a human population for the purpose of disease propagation:** Adapt the model of disease propagation developed in Sections 6.2 and 6.3 to represent coordination numbers which are *less* than 8. Populations where the average person only comes into close enough proximity to pass a disease with less than 8 people, could be incorporated into an extended model on the basis of random numbers. For example, if the coordination number is 4 then the probability for infection of any of the 8 neighbouring people could be taken as $4/8 = 0.5$. Illustrate how choosing whether a person is infected or not, affects the propagation of the disease. By performing sufficient simulations, deduce the expected proportion of infected people as a function of the immunisation rate x.

6.2 **Very sociable societies:** Show how the percolation theory of disease propagation could be adapted to represent societies where each person, on average, comes into close enough proximity to pass a disease, with *more* than 8 people.

6.3 **Immunisation rate development:** Develop the original disease propagation simulations, as presented in Section 6.3, to calculate the number of infected people, after the disease propagates fully, for an arbitrary number of simulations and for many different immunisation rates x. Demonstrate the validity of the program by repeating some of the calculations in Section 6.4.

6.4 **Critical density in multiple source systems:** Extend the simulations in Project 6.3 to multiple sources and deduce the critical densities of immunisation for 2, 3 and 4 sources of infection.

6.5 **Effect of coordination number on the mixed magnetic phases:** Under the restriction of a two-dimensional square array representing the crystal lattice, the natural coordination number (the number of nearest neighbours that any one atom has) is 8. Adapt the simulation developed in Task 6.3 along the lines of Project 6.1 and 6.2, to show the effect of coordination numbers of 6, 10 and 12.

6.6 **Percolation theories of other systems:** In this short introductory text, just two examples of the application of percolation theory have been introduced. Spend some time considering the world around you, and think of a system that could benefit from percolation theory simulations. Develop that simulation and show some aspect of the behaviour of the system.

7

Evolutionary Methods

7.1 GENETIC ALGORITHMS

Many classes of physical or scientific problems involve optimising a functional form in order to obtain the true solution. The variational method discussed in Chapter 2 is one example of this. The functional form of the solution is deduced from symmetry arguments, or intuition, or in desperate cases, a wild guess is made, and one or more parameters within the function are varied in order to minimise the energy of the system.

In some classes of problem many more parameters are required. For example, the solution of a quantum potential by expanding the wave function as a linear combination of basis states, is usually manipulated into a matrix form, as in Chapter 3, and then solved by direct diagonalisation. However, it could be recast into a variational calculation with the unknown coefficients treated as parameters. It has already been mentioned that the matrix, particularly in three-dimensional potentials, can get very large and can soon outgrow the memory available on today's computers. Thus, recasting the problem into an alternative form, might avoid this issue. Exploring a 100 dimensional parameter space can get tedious, even for a computer. However, computational methods which mirror evolution can offer a quicker route to optimise the solution.

The evolutionary approach, sometimes called a 'genetic algorithm', treats the variational parameters as 'genes' and an entire solution as a 'genome'. In a quantum mechanical problem, the genes composing the entire genome, could be the values of the wave function across spatial coordinates, or more simply they could just be the expansion coefficients for a wave function expressed as a linear series (for example, the a_j in equation (3.21)).

After randomly generating a set of potential solutions, the genomes are subjected to computational analogies of natural selection, reproduction and mutation over several generations, in the hope of evolving solutions which are close to the optimal, faster and by using less resources than are used in the alternative methods.

7.2 THE FITNESS FUNCTION

So, given a genome, a decision has to be made on how optimal, or how 'fit', it is as a solution. In the case of a racing car travelling around a track, then the optimum solution is clearly the one with the lowest time. For the salesman travelling from city to city, the optimum solution might be chosen in terms of the distance travelled. For quantum systems the 'fitness' of the genome would be, very simply, measured in terms of its *energy*. As discussed earlier in Section 2.6, the variational principle makes it clear that the lower the energy of a potential solution, the closer it is to the true solution of the system. Thus, for quantum, or classical, mechanical systems, the 'fitness function' could be just the energy.

The time-independent Schrödinger's equation reads:

$$\mathcal{H}\Psi = E\Psi \tag{7.1}$$

Thus multiplying on the left by Ψ^* and integrating over all space:

$$\int_{\text{all space}} \Psi^* \mathcal{H}\Psi \ d\tau = \int_{\text{all space}} \Psi^* E\Psi \ d\tau \tag{7.2}$$

but the energy E is a constant, so it can be brought outside of the integral, thus giving:

$$\int_{\text{all space}} \Psi^* \mathcal{H}\Psi \ d\tau = E \int_{\text{all space}} \Psi^* \Psi \ d\tau \tag{7.3}$$

Therefore:

$$E = \frac{\int_{\text{all space}} \Psi^* \mathcal{H}\Psi \ d\tau}{\int_{\text{all space}} \Psi^* \Psi \ d\tau} \tag{7.4}$$

or in the more compact notation of Dirac:

Fitness Function of a Quantum System

$$E = \frac{\langle \Psi | \mathcal{H} | \Psi \rangle}{\langle \Psi | \Psi \rangle} \tag{7.5}$$

Thus, given a genome, no matter how it was chosen, there is a well-defined procedure for evaluating its associated energy and hence for deciding its fitness.

7.3 RANDOM (MONTE CARLO) GENERATION OF GENOMES

Consider the expansion of the wave function that was introduced in the discussions on 'Matrix Methods' in Section 3.2:

$$\Psi = \sum_{j=1}^{n} a_j \psi_j \tag{7.6}$$

and assume also the corresponding basis states introduced in Section 3.1:

$$\psi_j = \sqrt{\frac{2}{L}} \sin\left(\frac{j\pi z}{L}\right), \quad j \in \mathcal{Z} \tag{7.7}$$

In an evolutionary approach, the wave function Ψ would be referred to as a potential solution, or genome, and the unknown coefficients a_j would be the genes. An evolutionary approach can be looked upon as just a way of by-passing the construction and subsequent solution of the Hamiltonian matrix ($\mathcal{M}_{ij}$).

Consider a simple computer code just to generate random sets (to a specified number of terms) of coefficients a_j and the subsequent evaluation of the fitness of the genome (the energy of the guessed solution).

```c
#include <stdlib.h>
#include <math.h>

#define hbar 1.05459e-34
#define m 9.109534e-31
#define e_0 1.602189e-19
#define pi 3.141593
#define L 200e-10        /* length of z domain */
#define N 10000          /* generate N random solutions */
#define Nb 4             /* number of basis functions */
#define Nz 1000          /* number of z coordinates */
#define seed 1           /* random number seed */

main()
{
float    a[Nb];          /* expansion coefficients (genes) aj */
float    Energy();       /* function for calculating the energy */
float    E;              /* the energy E (meV) */
int      iN;             /* index over N repetitions */
int      j;              /* index over columns */

srand(seed);             /* initialise random number sequence */


for(iN=0;iN<N;iN++)      /* repeat N times */
{
  for(j=0;j<Nb;j++)      /* Generate each gene randomly */
    a[j]=((float)(rand()-RAND_MAX/2))/(RAND_MAX/2);

  E=Energy(a);           /* evaluate the energy (fitness) */
  printf("%i %f\n",iN,E/(1e-3*e_0));
}

}  /* end main, function definitions below this line */
```

It can be seen from the code that the main() function is quite simple and meets all the objectives required. The array a[], consisting of Nb elements, is used to store the coefficients a_j, which are chosen randomly with the function rand() and then shifted in order to force them to lie within the limits $-1 < a_j < +1$. Note that only one set of coefficients, one genome, is stored at any one time. The energy E is calculated for each genome and printed, with the whole process being repeated N times.

The vast majority of the C-code, in the form of the functions below, is just dedicated to the calculation of the expectation value of the energy E for a given randomly generated wave function. Note that the nominator and denominator in equation (7.4) (referred to as top and bot) in the function Energy(), are calculated numerically. Some effort could have been made to deduce analytical expressions, but the numerical approach retains the powerful attribute of *generality*. In this example, the same parabolic potential as was used in Section 3.6 for the matrix approach, was employed. This choice was made in order to provide a ready-made example to compare with. This code can be adapted to different potentials, simply by changing the potential V() —there are no definite integrals to perform analytically.

Function
for
Expectation
Value of
Energy
(the Fitness)

```
float
Energy(a)

float    a[];
{
 float    d2Psi(); /* 2nd derivative of Psi */
 float    Psi();    /* the full wave function */
 float    V();      /* the potential V(z) */
 float    dz;       /* the integration step length */
 float    top=0;    /* <Psi|H|Psi> */
 float    bot=0;    /* <Psi|Psi> */
 float    z;        /* the spatial coordinate */
 int      iz;       /* index over z coordinates */

 dz=L/((float)Nz);

 for(iz=0;iz<Nz;iz++)
 {
  z=((float)iz)*dz;
  top+=Psi(a,z)*(-(hbar/(2*m))*hbar*d2Psi(a,dz,z)
                    +V(z)*Psi(a,z));
  bot+=Psi(a,z)*Psi(a,z);
 }

 return(top/bot);
}
```

```
float
Psi(a,z)              /* calculates the full wave function */


float    a[];
float    z;
{
 float   psi();    /* the basis functions */
 float   sum=0;    /* temporary value of Psi */
 int     j;        /* the index over the basis states */


 for(j=0;j<Nb;j++)
 {
  sum+=a[j]*psi(z,j);
 }
 return(sum);
}



float
d2Psi(a,dz,z)      /* calculates the 2nd derivative of Psi */


float    a[];
float    dz;
float    z;
{
 float   Psi();    /* the basis functions */


 return((Psi(a,z+dz)-2*Psi(a,z)+Psi(a,z-dz))/(dz*dz));
}



float
psi(z,k)              /*calculates basis function */


float    z;
int      k;
{
 /* Note the (k+1): lowest basis state is sin(1*pi*z/L) */
 return(sqrt(2/L)*sin((float)(k+1)*pi*z/L));
}
```

The Total Wave Function Ψ

Second Derivative of Wave Function

Basis State ψ_j

Potential
V(z)

```
float
V(z)                         /* returns the potential V(z) */

float    z;
{
  float   v;       /* the value of the potential */
  float   z0;      /* the barrier width */

  z0=L/2;

  v=((z-z0)/(100e-10))*((z-z0)/(100e-10))*0.1*e_0;

  return(v);
}
```

If many simulations are to be performed for a particular potential, then it might be worthwhile investing time in deducing more analytical expressions for the energy, but in these demonstrator cases, attention will be restricted to a numerical evaluation. Note the approach to this, the guessed wave function, or genome, Ψ is evaluated as a simple sum (see the function Psi()) and the second derivative is evaluated from a finite difference approximation (see the function d2Psi())—so effort doesn't even have to be spent in differentiating. The usual one-dimensional Hamiltonian has been assumed:

$$\mathcal{H} = -\frac{\hbar^2}{2m}\frac{\mathrm{d}^2}{\mathrm{d}z^2} + V(z) \qquad (7.8)$$

Fig. 7.1 shows a scatter plot of the energy for 100 randomly generated genomes, each with 4 genes, as calculated by the computer program above. It can be seen that they cover a range of energies from nearly 60 to less than 10 meV. The horizontal dashed line shows the result (6.95 meV) of the corresponding matrix approach with 4 basis states, as described previously in Table 3.1 of Section 3.6. The matrix method does produce the optimal solution, *for a given number of basis states*, thus randomly choosing the coefficients will at best only equal the 6.95 meV.

One hundred randomly generated genomes is a relatively small number and ideally the user might want to look at many more genomes. To aid this, the computer program was modified in order to print out the energy only if it was less than the previous minimum. Fig. 7.2 shows the results of this process.

It can be seen from the figure that, with 4 genes (basis states), the 'convergence' to an energy close to the optimum of 6.95 is relatively quick, requiring only around 20,000 'guesses'. However, when 6 genes are included, although the final result may be better, many more randomly generated genomes are required in order to produce energies comparable to the simulations with 4 genes.

The reason for this is that, if each gene a_j has to be within 10% of its optimum value say, in order to produce an energy which is close to the optimum, then if the problem requires N_b genes, the probability that this is chosen at random would be $(0.1)^{N_b}$. This would mean that on average $1/(0.1)^{N_b} = 10^{N_b}$ genomes would have to be created in order to have an evens (50%) chance of generating the correct one. The more genes, the more

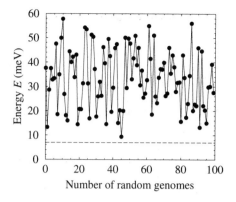

Fig. 7.1 A scatter plot of the energy E of 100 randomly generated genomes. The horizontal dashed line gives the lowest possible energy

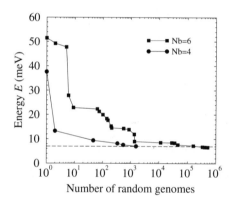

Fig. 7.2 The minimum value of the energy as a function of the number of randomly chosen genomes, for Nb=4 and Nb=6 basis states

genomes needed. For example, if the system requires 4 basis states in a matrix expansion, as in the simulations above, then it will require 4 genes and hence the expected number of randomly chosen genomes would have to be $10^4 = 10,000$. If 6 genes were required then the expected number of genomes would be $10^6 = 1,000,000$. Both of these numbers are commensurate with the results in Fig. 7.2.

Many real world problems, such as the solution of Schrödinger's equation in three dimensions require many more than 4 or 6 basis functions, thus this simple method of generating large numbers of potential solutions and selecting the best one on the basis of its fitness, is just not practical. Evolutionary methods or genetic algorithms are techniques by which these randomly chosen solutions can be combined in a systematic way in order to converge towards the optimum.

7.4 REPRODUCTION AND NATURAL SELECTION

Evolution is such a powerful process that it has even created a process by which it (evolution) can proceed quicker and more efficiently—*sexual reproduction*.

Sexual reproduction is incredibly powerful because it produces a 50:50 mixture of genes, from two parents, in one offspring, in just one generation—a very short time on evolutionary timescales. What's more the mixture is different each time (ignoring identical twins), so two parents can produce several (or many) offspring with quite different traits. Now, these different characteristics might seem very small, such as a 3 cm difference in height or a difference in strength or resistance against disease, however, there may be some characteristic which could lead to the offspring having a marginally better chance of surviving into adulthood and reproducing themselves. Thus, as evolution goes, genes which help an individual to survive and reproduce are more likely themselves to be passed on and survive within the individual's descendants[1].

Evolution in the context of numerical methods, is represented by the evolution of solutions. In a population of solutions that make up a generation, 10% say, of the fittest, could be allowed to survive and reproduce in the hope that some of their offspring might be even fitter. Thus, over the course of many generations, a very fit, or optimum, solution might emerge. Fig. 7.3 attempts to represent this schematically.

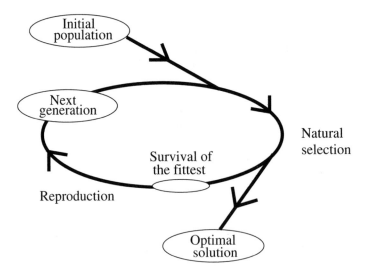

Fig. 7.3 The evolutionary cycle: After the initial creation of a random population (set of solutions), 10% of the fittest are allowed to survive and reproduce. This new generation is then subject to the same fitness selection and so on

---

[1]Often called 'survival of the fittest'. However, one could look upon it from the gene's point of view—perhaps they are very selfish and the driving force is the survival of the genes and not the genome it makes up.

The C-code below shows one way that the program in the last section, which just randomly generated solutions, could be adapted to continuously update a list of the fittest genomes.

```
#include <stdlib.h>                                      C Code
#include <math.h>                                        to Select
                                                         Fittest
                                                         Genomes

#define hbar 1.05459e-34
#define m 9.109534e-31
#define e_0 1.602189e-19
#define pi 3.141593
#define L 200e-10        /* length of z domain */
#define BIG 1            /* large energy, useful initial condition */
#define N 100            /* generate N random solutions */
#define Nb 4             /* number of basis functions */
#define Nr 10            /* number allowed to reproduce (Nr<Ng) */
#define Nz 1000          /* number of z coordinates */
#define seed 2           /* random number seed */


typedef struct{          /* define a structure to hold both the */
float    a[Nb];          /* coefficients aj and the corresponding*/
float    E;              /* fitness (energy) */
}genome;


main()
{
genome   any;            /* ANY old genome */
genome   fittest[Nr];    /* the genomes that will reproduce */
float    Energy();       /* function for calculating the energy */
int      iN;             /* index over N repetitions */
int      ir;             /* index over Nr fittest genomes */
int      iR;             /* index over Nr fittest genomes */
int      j;              /* index over columns */


srand(seed);             /* initialise random number sequence */

                                                         Initialise
for(ir=0;ir<Nr;ir++)     /* initialise fittest as very unfit */   List of
  fittest[ir].E=BIG;                                     Fittest
```

Generate
N Random
Genomes

```
for(iN=0;iN<N;iN++)          /* repeat N times */
{
  for(j=0;j<Nb;j++)          /* Generate each gene randomly */
    any.a[j]=((float)(rand()-RAND_MAX/2))/(RAND_MAX/2);

  any.E=Energy(any.a);       /* evaluate the energy (fitness) */
  for(ir=0;ir<Nr;ir++)       /* compare with the Nr fittest */
  {
    if(any.E<fittest[ir].E)            /* if fitter than any in list */
    {
      for(iR=Nr-2;iR>=ir;iR--)         /* consider all those higher */
      {
        fittest[iR+1]=fittest[iR];   /* and move them up */
      }
      fittest[ir]=any;         /* insert new fit genome */
      any.E=BIG;               /* make BIG to avoid repeated insertion */
    }
  } /* end for ir */
} /* end for iN */
```

Compare
Energy with
Fittest

If 'Fit'
Insert
in List

Print
List of
Fittest
Genomes

```
for(ir=0;ir<Nr;ir++)       /* Now output the fittest */
{
  printf("%f",fittest[ir].E/(1e-3*e_0));
  for(j=0;j<Nb;j++)printf(" %f",fittest[ir].a[j]);
  printf("\n");
}

}  /* end main, function definitions below this line */
```

As it is now necessary to keep a record of the genes making up each genome, a structure has been defined at the beginning of this code which stores the coefficients a_j in an array a[j] and the fitness (energy) E. The program defines a list of the Nr fittest[] genomes and initially makes them very unfit by setting their energy to a BIG value.

After generating the genomes randomly, as before, the program now compares them to each of the Nr fittest[] genomes. If the current (any) genome is fitter, then it moves all the other genomes up the list and slots the current genome into its rightful place. Note that all the functions for evaluating the energy of each genome remain unchanged from the previous section and are not included here. Finally, the program prints out the energy and coefficients a_j constituting the fittest genomes.

Table 7.1 shows the results of two simulations, each producing 100 randomly generated genomes, and selecting the 10 fittest, but the first with the random number sequence seed set at 1 and the second with it set at 2.

It can be seen from the table that each simulation only manages to produce an energy $\sim$9 meV in 100 guesses, compared to the known best value of 6.95 meV. However, what is

Table 7.1 The 10 fittest genomes produced in two consecutive simulations

seed=1				
E (meV)	a_1	a_2	a_3	a_4
9.366462	-0.281809	0.104970	0.158860	-0.094848
13.253349	-0.511346	-0.347973	0.458720	0.277308
13.282781	0.823295	-0.604897	-0.329554	0.536459
14.415075	-0.431413	0.477069	0.279958	-0.291903
14.528957	-0.647579	-0.519875	0.595596	0.465309
14.981381	-0.901675	-0.101790	0.972934	0.415818
15.026192	-0.991677	0.846138	0.187784	-0.639255
15.616261	0.676267	0.448504	-0.643585	-0.556069
15.981101	-0.615572	0.326454	0.780465	-0.302214
16.842909	-0.704680	0.762124	0.282161	-0.136093

seed=2				
E (meV)	a_1	a_2	a_3	a_4
9.414765	0.782832	-0.342481	-0.483528	0.173409
10.502014	0.780331	-0.337190	-0.604781	0.229484
13.499647	-0.953845	0.744707	0.843150	-0.271556
14.285234	-0.776645	0.341043	0.883677	-0.296275
15.015240	0.826593	0.091326	-0.934802	-0.340443
15.250257	-0.794064	-0.910199	0.289982	0.607889
15.587286	0.959742	0.801020	-0.617725	-0.007526
15.956071	0.859388	-0.909282	-0.421474	0.186701
15.971031	-0.805897	0.629854	0.565747	0.050237
16.191734	0.421350	0.317595	-0.251870	0.041969

more interesting can be seen by comparing two different genomes which have very similar energies, such as the first one in each list (9.37 and 9.41 meV) or eighth and ninth in the second list (15.96 and 15.97 meV). Although the energies of these trial wave functions are very similar, their expansion coefficients a_j are completely different, not in the least bit similar. Again this highlights the point made earlier in Chapter 2—in the context of the variational principle—the energy is a rather insensitive measure of the wave function.

Thus, on the road to developing a simulation based on evolutionary principles, a program has now been constructed which can generate an initial large population and select and order the fittest individuals. It remains now to breed...

7.5 IMPLEMENTATION OF A GENETIC ALGORITHM

Thus a random population has been generated and the fittest individuals can be selected and allowed to combine to produce, what it is hoped will be, even fitter solutions. There are many ways that reproduction could be summarised algorithmically, for example:

- Any one genome (the mother, for example) combines with just one other genome (the father), and each individual gene within the child genome is an exact replica of the same gene from either of the parents.

- As previously, but each gene is now some mixture of the parents' genes.

- Each genome is allowed to reproduce with many other genomes but, again, with any one gene being a replica from either of the parents.

- Each genome is allowed to reproduce with many other genomes and each gene is a mixture of the parents' genes.

At this point, attention will be focussed on the first of these and is illustrated in the C function 'mate' below:

A Mating
Function

```
genome                      /* combines two genomes to */
mate(mother,father)         /* produce a child */

genome   mother;
genome   father;
{
  genome child;
  int    j;                 /* index over genes */

  for(j=0;j<Nb;j++)         /* consider each gene in turn */
  {
    if(((float)rand()/RAND_MAX)<0.5)
      child.a[j]=mother.a[j];
    else
      child.a[j]=father.a[j];
  }

  return(child);
}
```

Note the function is of the type genome. Its input is two genomes, which may have been chosen randomly or systematically, and for the purposes of the reproduction the function labels them as 'mother' and 'father', these are just convenient labels and could just as well have been 'pen' and 'cob'. The purpose of the function is to produce a new genome, so it considers each gene of the offspring in turn and assigns that gene as either the gene of the mother or the gene of the father depending on the value of a random number. Thus the resultant combination has a stochastic component and each particular pair of parent genomes can produce 2^{Nb} *distinct* genomes.

Building on the codes developed so far in this chapter, the C-code below shows how this function can be incorporated in order to produce a full evolutionary simulation—a genetic algorithmic solution:

```
#include <stdlib.h>
#include <math.h>

#define hbar 1.05459e-34
#define m 9.109534e-31
#define e_0 1.602189e-19
#define pi 3.141593
#define L 200e-10          /* length of z domain */
#define BIG 1              /* large energy, useful initial condition */
#define N 100              /* number of genomes in a generation */
#define Nb 6               /* number of basis functions */
#define Ng 4               /* number of generations */
#define Nr 10              /* number allowed to reproduce (Nr<N) */
#define Nz 1000            /* number of z coordinates */
#define seed 1             /* random number seed */

typedef struct{            /* define a structure to hold both the */
float    a[Nb];            /* coefficients aj and the corresponding*/
float    E;                /* fitness (energy) */
}genome;

main()
{
float    Energy();         /* function for calculating the energy */
genome   mate();           /* mates to genomes to produce child */
void     order();          /* arrange fittest[Nr] genomes in order */
void     printgenome();    /* prints a genome to screen */
genome   any;              /* ANY old genome */
genome   child[N];         /* children genomes */
genome   fittest[Nr];      /* the genomes that will reproduce */
int      ic;               /* index over children */
int      ig;               /* index over generations */
int      iN;               /* index over N genomes */
int      ir;               /* index over Nr fittest genomes */
int      j;                /* index over columns */

srand(seed);               /* initialise random number sequence */

for(ir=0;ir<Nr;ir++)       /* initialise fittest as very unfit */
  fittest[ir].E=BIG;
```

C Code for Genetic Algorithm

```
for(iN=0;iN<N;iN++)          /* generate N initial genomes randomly */
{
  for(j=0;j<Nb;j++)          /* Generate each gene randomly */
    any.a[j]=((float)(rand()-RAND_MAX/2))/(RAND_MAX/2);
```

Maintaining
List of
Fittest
now in
Function

```
  any.E=Energy(any.a);       /* evaluate the energy (fitness) */
  order(any,fittest);        /* keep cumulative record of Nr fittest */
}/* end for iN */

printgenome(fittest[0]);/* output fittest genome */
```

Subsequent
Generations

```
for(ig=0;ig<Ng;ig++)         /* repeat for Ng generations */
{
  iN=0;                      /* set counter for children to zero */
  for(ir=0;ir<(Nr/2);ir++)          /* mate half with other half */
  {
    for(ic=0;ic<(2*N/Nr);ic++)
```

Create
Children

```
    {
      child[iN]=mate(fittest[ir],fittest[Nr-ir-1]);/* monogamy */
      child[iN].E=Energy(child[iN].a);      /* evaluate fitness */
      iN++;                  /* increment counter for children */
    }
  }
```

Maintain
List of
Fittest

```
  for(iN=0;iN<N;iN++)
    order(child[iN],fittest);

  printgenome(fittest[0]);/* output fittest genome */
}
```

```
} /* end main, function definitions below this line */
```

The preamble to the main() function is the same as previously, with the exception of the addition of the definition of the number of generations (iterations) required in the simulation. Note also the introduction of the array of genomes which has been called child[], this will be used to store all N genomes in each generation.

N genomes are again created randomly to begin the simulation. However, now the algorithm to order the fittest individuals has been moved to a function to allow for easy reuse:

Fittest
List
Ordering
Function

```
void
order(any,fittest)           /* orders fittest genomes */

genome  any;
genome  fittest[];
{
  int   ir;                  /* index over Nr fittest genomes */
  int   iR;                  /* index over Nr fittest genomes */
```

```
for(ir=0;ir<Nr;ir++)      /* compare with the Nr fittest */
{
  if(any.E<fittest[ir].E)          /* if fitter than any in list */
  {
    for(iR=Nr-2;iR>=ir;iR--)       /* consider all those higher */
    {
      fittest[iR+1]=fittest[iR];   /* and move them up */
    }
    fittest[ir]=any;         /* insert new fit genome */
    any.E=BIG;               /* make BIG to avoid repeated insertion */
  }
} /* end for ir */

}
```

After creating the initial population and ordering the Nr fittest as before, the main function then prints out the fittest (one) genome. Again, the code to do this has been hived off into a function printgenome to allow for easy reuse:

```
void                                              Genome
printgenome(any)                                  Printing
                                                  Function
genome   any;
{
  int    j;          /* index over genes */


  printf("%f",any.E/(1e-3*e_0));           /* print energy */
  for(j=0;j<Nb;j++)printf(" %f",any.a[j]); /* genes */
  printf("\n");                            /* end line */

}
```

Referring back to the main() function above, the genetic algorithm then proceeds by opening a loop for Ng iterations. Within this loop the irth fittest genome is mated with the Nr-1-irth genome[2]. This is repeated to produce 2*N/Nr new genomes—some of which may be fitter than their parents. This number was chosen because Nr reproducing genomes produce Nr/2 pairings, and with the above number of children per family, the full population of N can be produced for each generation.

After the population has been replenished, the order function updates the list of fittest genomes produced so far (over all previous generations). The current fittest genome is output, and the next generation is created.

---

[2]For example, if 10 genomes were allowed to reproduce, then this algorithm would mate the 1st fittest (fittest[0]) with the 10th fittest (fittest[9]) and then the 2nd fittest with the 9th, etc. There is no real reason for this choice, it could have been the 1st with 2nd, 3rd with 4th, etc.

The first two columns of Table 7.2 just tabulate the data given earlier in Fig. 7.2, i.e. the current minimum energy versus the number of randomly generated solutions. The last three columns, however, display the results of executing the genetic algorithm developed above for Nb=4 and 6 basis states (genes). Focusing on the 4 gene genomes first, it can be seen that the genetic algorithm converges to an energy of 7.512 meV after considering 200 possible solutions. However, after the random approach has considered 200 solutions, the minimum energy that it has achieved is 9.366 meV, and indeed the random technique requires 1543 guessed solutions before it can beat that produced by the genetic algorithm in 200. A similar pattern can also be observed for the larger simulation with 6 genes.

Despite the relatively small sample set and the uncertainties inherent with any stochastic analysis, there *are* some efficiency savings in attempting to optimise randomly generated solutions with a genetic algorithm. However, the solutions are not optimal: evolution does drive towards lower energies and closer approximations to the exact solution, but there is still some way to go—*there is a limitation in the simulation.*

7.6 MUTATION AND CONCLUSION

The limitation in the evolutionary method developed so far is that the simulation is restricted to using the initial 'gene pool'—the genes that are randomly generated for the first generation. Thus if the only hair genes created at the beginning were 'brown' and 'black', then all subsequent offspring would have either 'brown' or 'black' hair. Nature isn't like this: the genes themselves do change. They may do this by mixing, so that a 'brown' hair gene and a 'black' hair gene combine and produce a 'blonde' hair gene. It is possible to represent this, say by averaging the values the genes have—or perhaps by taking a weighted average according to the fitness of the genome—the subject of Task 7.2.

In addition, genes 'mutate'—or spontaneously change. This may occur because of a change in the chemical makeup, perhaps because of a direct hit by a cosmic ray, or perhaps for some other reason, but it does happen. Such a process could be included in an evolutionary method by occasionally changing the value of a gene, for example a randomly chosen gene in every tenth genome could be given a completely new value determined by a random number, or whatever—the implementation of this is the subject of Project 7.2.

It has been demonstrated that, given some thought, the majority of mechanisms driving the evolution of species can be implemented in a computational model which drives a collection of solutions towards an optimum. Such simulations offer greater efficiency in classes of physical and mathematical problems where the sought-after solution has to be represented by a linear combination of basis functions. Indeed, such an evolutionary method avoids the need to set up a complete non-sparse matrix (which can very quickly outgrow available computer resources) for direct diagonalisation.

Table 7.2 A comparison of the data from the random solutions (first two columns) with that from the genetic algorithm (last three columns). N is the number of randomly generated solutions with the corresponding minimum energy E, and Ng is the number of generations. As each generation consists of 100 potential solutions, then the total number of solutions considered is 100Ng, as given in the fourth column

N	E (meV)	Ng	100Ng	E (meV)
		Nb=4		
0	37.607042	1	100	9.366462
1	13.282781	2	200	7.512481
45	9.366462	3	300	7.512481
336	8.195581			
539	7.612535			
1543	6.997562			
		Nb=6		
0	51.468596	1	100	21.239543
1	49.190035	2	200	15.467397
4	47.723054	3	300	13.294649
5	27.874318	4	400	12.193126
8	22.882695	5	500	12.193126
63	22.335187			
73	21.239543			
100	19.923905			
138	18.306291			
152	17.649503			
197	15.244999			
209	14.508963			
557	14.282856			
864	13.817770			
1346	11.870637			

Summary

It has been demonstrated that the evolutionary laws of nature of natural selection, survival of the fittest, reproduction and mutation (see projects below) can be applied in a computational sense to optimise a solution to a physical problem. In particular, the natural selection of the fittest potential solutions (or genomes) can be performed on the basis of a determinable physical quality, such as the time or length, or in the case of quantum mechanical systems, the energy:

$$E = \frac{\langle \Psi | \mathcal{H} | \Psi \rangle}{\langle \Psi | \Psi \rangle}$$

Tasks

7.1 **Different selection ratios:** Investigate the effect of altering the proportion of a population allowed to reproduce. Use the simulation and example above.

7.2 **Gene mixing:** Implement the suggestion in Section 7.5 of mixing the genes during the reproductive mechanism. Illustrate the effect this has by repeating the example simulation of the parabolic quantum well with 4 basis states (genes).

7.3 **Real space optimisation of parabolic potential solutions:** Adapt the simulation in Section 7.5 to consider a system of genomes where the genes are just the values of the wave function in real space at 10 points along a one-dimensional axis, i.e. $\psi(z)$. Using suitably chosen computational parameters and the parabolic potential as previously find the optimum wave function.

Projects

7.1 **Implement polygamous genome reproduction:** Remove the restriction of only allowing one genome to reproduce with just one other genome, and investigate the effect it has on the example simulation.

7.2 **Implement gene mutation:** After producing the next generation of solutions by reproduction, mutate some of the genes within some of the genomes (by either altering the value of a gene by some amount, or by swapping the positions of two genes within a genome). Investigate the effect that the amount of mutation has on the efficiency of the simulation.

7.3 **Non-stochastic optimisation using the steepest descent method:** An alternative method for optimising a potential solution formulated as a Fourier-type series is to use the non-stochastic method of 'steepest descent'. In this technique the *change* in the associated energy of a potential solution is evaluated for small (positive and negative) independent changes in each coefficient. The solution which produces the largest reduction in the energy is taken as the next iterated solution and the procedure is repeated. Demonstrate this method using the same example of the parabolic quantum well and compare the efficiency of this method with the stochastic genetic algorithm.

8

Molecular Dynamics

8.1 MODELLING MOLECULES AND CRYSTALS

The electronic and pharmaceutical industries are perhaps the two most important contributors to the rapid development of human technologies over the last half-century. And they have one thing in common—they are both centred around the microscopic properties of materials. In the case of electronics, the properties of semiconductor crystals are fundamental to their exploitation in devices, and this is particularly true of the developments over the last twenty years or so, which have been based around semiconductor heterostructures[1].The pharmaceutical industry is driven by the desire to develop medicines (drugs), which are almost entirely derived from complicated organic molecules.

In both technologies the need to understand the materials at the microscopic (atomic or molecular) level is paramount. In the early days the molecules and crystal structures were reconstructed at the human level by making models consisting of wooden (and later plastic) balls connected together by rods or sticks. The balls represented the individual atoms within the chemical compound and the rods represented the bonds holding the atoms together. Such models helped the scientists to visualise the material in their own minds, and indeed to see where other molecules might be able to attach and hence form still more complex compounds.

The dawn of the computer allowed this model-making to be performed 'virtually', and in the case of molecules consisting of many hundreds of atoms, much more quickly. But more importantly, the computer allowed *numerical* models to be constructed which allowed the scientist to experiment with different shapes and alignments and to determine

---

[1] See, for example, M. J. Kelly, *'Low Dimensional Semiconductors: Materials, Physics, Technology, Devices'* (Clarendon, Oxford, 1995).

their properties. For example, which structure would have the lowest energy and hence be more likely to occur naturally. Such a technique has become known as 'Molecular dynamics' and it is now a very important tool which aids the development of many branches of science and technology which depend intimately on the properties of materials. In this chapter, a brief introduction to this method will be given and some simple algorithms will be developed which give an appreciation of the extreme usefulness of this technique.

8.2 ATOM–ATOM INTERACTIONS AND INTERACTION POTENTIALS

Consider two hydrogen atoms[2] in a vacuum separated by a reasonably large distance, say many atomic radii. Now it is known that hydrogen likes to bond in pairs of atoms to form hydrogen molecules H_2, hence the original hydrogen atoms must feel an attractive force. This is illustrated in the top part of Fig. 8.1.

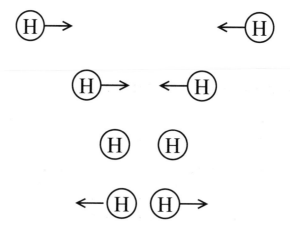

Fig. 8.1 The forces on a pair of hydrogen atoms (H) at various distances apart

This force will draw the hydrogen atoms together until eventually they reach an equilibrium distance and bond to form a molecule. At this distance, as given by the third of the illustrations in Fig. 8.1, the potential energy of the pair of hydrogen atoms is at a *minimum*, any further reductions in separation of the atoms will be resisted because of the mutual repulsion of the positively charged atomic nuclei. This is known from experiment, it is known that the formation of hydrogen molecules from hydrogen atoms is a reversible process, and that there is a definite 'bond length' describing the distance between the centres of the two atoms. The two hydrogen atoms do not keep attracting each other and disappear into one 'ball' of matter, thus there definitely is a repulsive force which acts at

---

[2]The hydrogen atom consists of a single positively charged proton, orbited by a single negatively charged electron at a Bohr radius of 0.528 Å, recall Sections 1.12 and 2.7. The 'electronic structure' is therefore summarised as $1s^1$, which means the 1s orbital contains 1 electron.

small distances. This interaction could be described in terms of the potential energy of the system and consideration of the arguments presented above, allows it to be illustrated qualitatively, as in Fig. 8.2.

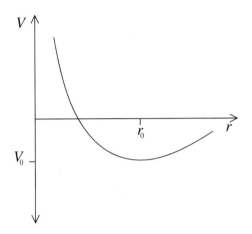

Fig. 8.2 The potential energy V of two hydrogen atoms separated by a distance r. The potential energy minimum V_0 represents the bonding energy of a hydrogen molecule and the equilibrium distance r_0 the bond length

There are many mathematical forms for the potential energy curve in Fig. 8.2, and great discussions on how to represent the various types of chemical bonding such as covalent, ionic, van der Waals and hydrogen bonding—see any good solid state physics book[3].

In this work, the choice of Callister[4] will be followed. Not for any physical reason, merely that the potential energy is given by a simple mathematical expression that can be easily coded—because this book is concerned with the illustration of computational methods and not the details of particular chemical compounds. And in fact, the above source quotes the potential energy as:

$$V = -\frac{A}{r^m} + \frac{B}{r^n} \qquad (8.1)$$

An Interaction Potential

where A, B, m and n are positive constants and chosen empirically to describe any particular chemical bond. The negative sign of the first term indicates that this represents the attractive component of the interaction between the atoms, and the index m is typically of order 1, in order to give the long-range attractive force that has been postulated. The second term has a positive sign and represents the repulsive force, the index n characterising the behaviour of this component is usually much greater than 1, thereby ensuring a short-range nature.

[3]For example, J. S. Blakemore, *'Solid State Physics'*, (University Press, Cambridge, 1985),
S. O. Kasap, *'Principles of Electrical Engineering Materials and Devices'*, (McGraw-Hill, Boston, 1997),
N. W. Ashcroft and N. D. Mermin, *'Solid State Physics'* (Saunders, Philadelphia, 1976).
[4]W. D. Callister, *'Materials Science and Engineering'* (Wiley, New York, 1985).

The force F itself can be found simply by differentiating the potential energy, which, as in other areas of physics, is given by:

$$F = -\frac{\partial V}{\partial r} \tag{8.2}$$

Thus:

$$F = -\frac{\partial V}{\partial r} = -\frac{\partial}{\partial r}\left(-Ar^{-m} + Br^{-n}\right) \tag{8.3}$$

*The
Resulting
Force*

$$\therefore F = -mAr^{-m-1} + nBr^{-n-1} \tag{8.4}$$

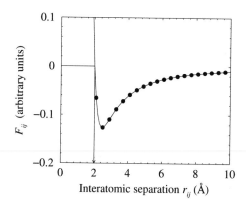

Fig. 8.3 The force F_{ij} between two atoms i and j, as a function of their separation r_{ij}, for the chosen equilibrium separation, r_0, equal to 2 Å.

The equilibrium distance r_0 is just the separation at which the force is zero, i.e.

*The
Equilibrium
Separation*

$$0 = -mAr_0^{-m-1} + nBr_0^{-n-1} \qquad \therefore r_0^{-m+n} = \frac{nB}{mA} \tag{8.5}$$

Thus if the indices m and n describing the attractive and repulsive forces are chosen first, a desired equilibrium distance r_0 can be obtained by choosing the ratio of the coefficients A to B using equation (8.5). This is illustrated in Fig. 8.3. The equilibrium distance r_0 was chosen as 2 Å and the arbitrary choice of the coefficient A as 1, allowed B to be chosen. It can be seen from the figure that this requirement was met, as the separation for zero force is indeed 2 Å[5]. The actual magnitude of the force is not too important in this work, hence the arbitrary choice of the coefficient A as 1. All that the magnitude of the force will influence is the rate at which the atoms will settle into their coexisting equilibrium, the only point of interest in this introduction are the equilibrium positions themselves[6].

---

[5]The computer code which generated this data is part of the full simulation developed in the next section. Note, the more specific labelling of the force as F_{ij} and the separation between two atoms as r_{ij} has been introduced for later referral.

[6]More realistic interatomic potentials, and their resultant interatomic forces, allow the evaluation of properties such as structural strength and heat capacity. And indeed, there is a whole field of research into such matters. However, this lies well beyond the scope of this introduction.

8.3 SIMULATIONS OF ATOMIC ENSEMBLES

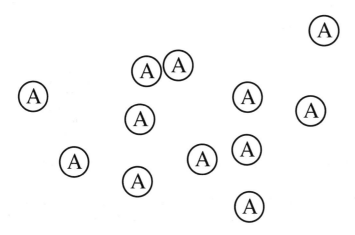

Fig. 8.4 An ensemble of identical atoms

A molecular dynamics simulation consists of a computational model of a system of atoms, the interactions between which are described by potentials as in Fig. 8.2. For the case of hydrogen, a simulation of two hydrogen atoms would be rather boring, because no matter what the initial conditions of the two atoms might be, there would only be two possible outcomes: (i) the atoms bond to form a molecule, or; (ii) the atoms' initial kinetic energy is so large that they have an 'escape velocity' and never bond.

However, consider now a system of many identical atoms, the interaction between any two of which can be described by a *spherically symmetric* potential of the form given in Fig. 8.2, with perhaps a different equilibrium distance r_0 and a different minimum potential V_0, see Fig. 8.4. It is obvious that the atoms will bond to form a solid, but what will be the crystal structure of that solid?

A simulation might proceed by randomly allocating the initial positions of a number, say N, of these atoms, and then allowing them to relax into positions which will minimise the total potential energy of the system. The latter represents an equilibrium situation and provides a well-defined end point. Perhaps the simplest method of determining which way an atom within the interacting system should move, would be to calculate the total force upon it and move the atom a small distance in the direction of the force, see illustration in Fig. 8.5.

The force on the ith atom is given by the sum of all the forces arising from the interactions with every other atom:

$$\mathbf{F}_i^{\text{total}} = \sum_{j=1, j \neq i}^{N} \mathbf{F}_{ij}(r_{ij})$$

(8.6)

The Total Force

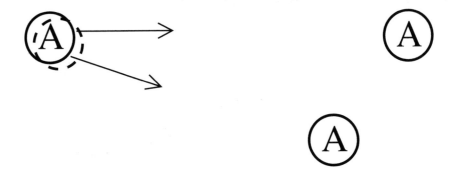

Fig. 8.5   The total force on any one atom and the resultant response within the simulation

The spherical symmetry is specified explicitly by the dependence of the interaction potential on the separation r_{ij} of the atoms only. It can be obtained directly from Pythagoras:

$$r_{ij}^2 = (x_j - x_i)^2 + (y_j - y_i)^2 + (z_j - z_i)^2 \qquad (8.7)$$

The vector nature of each interaction has to be accounted for, and this can be achieved geometrically as illustrated in Fig. 8.6. Using simple trigonometry, then the component of the force $\mathbf{F}_{ij}$ along the x-axis would be:

$$F_{ij}^x = |\mathbf{F}_{ij}| \cos \theta, \quad \text{where} \quad \cos \theta = \frac{x_j - x_i}{r_{ij}} \qquad (8.8)$$

and similarly, along the y-axis:

$$F_{ij}^y = |\mathbf{F}_{ij}| \sin \theta, \quad \text{where} \quad \sin \theta = \frac{y_j - y_i}{r_{ij}} \qquad (8.9)$$

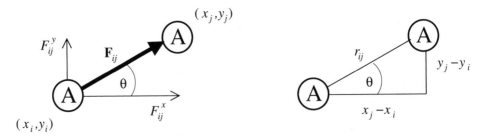

Fig. 8.6   Resolving the components of the interatomic force F_{ij} along the Cartesian axes (in this two-dimensional example, x- and y- only)

One implementation of the simulation would be to consider each atom i in turn, and calculate the net force upon it by summing the contributions over all other atoms j. Once this is complete, the atom's response to the force could be to allow it to move in the direction of the force by an amount proportional to the magnitude of the force. This would be repeated until an equilibrium is reached when the net force on each atom is zero. The algorithm describing this procedure is Algorithm 8.1.

Algorithm 8.1 Algorithmic description of a simple molecular dynamic simulation

$e \Leftarrow 1.60219 \times 10^{-19}$ {Define physical constants, for example}
$N \Leftarrow 12$ {Define system parameters, number of atoms}
$L \Leftarrow 30$ {Extent of initial distribution of atoms}
$\alpha \Leftarrow$ some-suitably-chosen-value
Generate randomly the atomic positions (x_i, y_i)
for I=1 to some-predefined-limit repeat **do**
    for i=1 to N **do**
        for j=1 to N, excluding i **do**
            Calculate r_{ij} {The distance from atom i to j}
            Calculate x- and y-components of F_{ij}
            Add F_{ij}^x and F_{ij}^y to cumulative sums F_i
        end for
        $x_i \Leftarrow x_i + \alpha F_i(x)$ {Change coordinates of atom i}
        $y_i \Leftarrow y_i + \alpha F_i(y)$
    end for
end for

In this initial example the atoms are to be restricted to movement in a plane, hence the omission of the z-coordinate in Algorithm 8.1. The only other point of note is the introduction of the 'relaxation constant' α, which is just a constant of proportionality linking the total force on an atom F_i, with its resulting displacement. A suitable value for α will be chosen by numerical experiment later.

8.4   COMPUTATIONAL IMPLEMENTATION

A C implementation of Algorithm 8.1 might look like:

```
#include <math.h>
#include <stdio.h>
#include <stdlib.h>
#define alpha 1          /* proportionality constant */
#define N 12             /* number of atoms */
#define L 10             /* size of system, in Angstrom */
#define r0 2             /* equilibrium separation, in A */
#define seed 1           /* random number sequence seed */




main()
{
double   dx;            /* movement of atom i along x */
double   dy;            /* movement of atom i along y */
double   Fij();         /* force between atoms i and j */
double   Fxi;           /* x-force on atom i */
double   Fyi;           /* y-force on atom i */
double   rij;           /* distance between atom i and j */
double   x[N];          /* x-coordinates of atoms, in A */
double   y[N];          /* y-coordinates of atoms, in A */
int      I;             /* index over Iterations */
int      i;             /* index over atoms */
int      j;             /* index over all other atoms */
char     filename[9];   /* name of output file */
FILE     *Fout;         /* pointer to output file */




srand(seed);            /* initialise random number sequence */




for(rij=0.1;rij<(4*r0);rij+=0.1)          /* output Fij(r) to */
  printf("%le %le\n",rij,Fij(rij));       /* screen */
```

Generate
Initial
(Random)
Atomic
Positions,
Save to File
```
Fout=fopen("xy0.r","w");
for(i=0;i<N;i++)
{
 x[i]=L*(float)rand()/RAND_MAX;           /* generate initial */
 y[i]=L*(float)rand()/RAND_MAX;           /* atomic positions */
 fprintf(Fout,"%le %le\n",x[i],y[i]);     /* write to file 'xy0.r' */
}
fclose(Fout);
```

```
for(I=1;I<1000;I++)          /* repeat many times */
{
 sprintf(filename,"xy%i.r",I);   /* new file each I */
 Fout=fopen(filename,"w");
 for(i=1;i<N;i++)             /* i=1 implies i=0 atom fixed */
 {
  Fxi=0;Fyi=0;               /* reset cumulative sums */
  for(j=0;j<N;j++)
  {
   if(j!=i)
   {
    rij=sqrt((x[j]-x[i])*(x[j]-x[i])+
             (y[j]-y[i])*(y[j]-y[i]));
    Fxi+=Fij(rij)*(x[j]-x[i])/rij;
    Fyi+=Fij(rij)*(y[j]-y[i])/rij;
   }
  }
  dx=alpha*Fxi/Fij(2*r0);          /* allow coordinates to alter */
  dy=alpha*Fyi/Fij(2*r0);          /* in proportion to force */
  if(dx<(-L/100.0))dx=-L/100.0;if(dx>(L/100.0))dx=L/100.0;
  if(dy<(-L/100.0))dy=-L/100.0;if(dy>(L/100.0))dy=L/100.0;
  x[i]+=dx;                       /* move atom */
  y[i]+=dy;                       /* move atom */
  fprintf(Fout,"%le %le\n",x[i],y[i]);   /* output positions */
 }
 fclose(Fout);
}

}/* end main */
```

Calculate Atomic Separations, Components of Force

Move Atom

Function definitions below this line

```
double
Fij(rij)          /* The interatomic force definition */

double   rij;
{
 double A=1;      /* coefficient of attractive term */
 double B;        /* coefficient of repulsive term */
 int     n=8;     /* index of repulsive potential term */
 int     m=1;     /* index of attractive potential term */

 B=m*A*pow(r0,-m+n)/n;    /* calculate B from defined r0 */

 return(-m*A*pow(rij,-m-1)+n*B*pow(rij,-n-1));
}
```

Interatomic Force Function

After the initialisation, the above code generates a list of data—F_{ij} versus r_{ij}—and writes it to the standard output (usually the screen). This is for diagnostic purposes (just to check the interatomic force looks as it should), the data can be easily captured and plotted, as done earlier in Fig. 8.3. The initial atomic positions are generated from random numbers and scaled to a two-dimensional box of side L.

As the method of convergence is of interest, and not just the initial and final positions, it was decided to output all the atomic positions at each step in the iterative procedure. The initial positions are immediately written to the file 'xy0.r', with subsequent positions, written at every iteration (as indexed by 'I') to files of the form 'xyI.r'.

Examples of the randomly generated *initial* positions of the atoms are displayed in Fig. 8.7, for four different random number seeds (1, 2, 3 and 4).

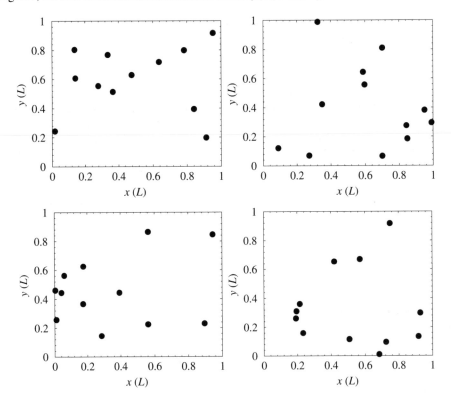

Fig. 8.7 Four example sets of randomly generated initial atomic positions, for N (the number of atoms) equal to 12

The code then largely follows the algorithm—calculating the total force on each atom in turn, and then allowing the atom to move a small distance in the direction of the force. There are three points of detail that had to be introduced, during prototyping this code, that are worthy of a mention:

1. It was found, during initial tests, that as the atoms converged into their equilibrium positions, they also drifted *en masse*. It is thought the reason for this arose from

the asymmetry of the initial random distribution, such that the 'centre of gravity' was not fixed upon any one atom. The solution was to fix the position of one of the atoms. This was achieved by increasing the initial value of the loop index i from 0 to 1, thus atom i=0 was not allowed to alter its position in response to the forces upon it.

2. In this introduction to molecular dynamics, a very simple expression has been employed for the interaction potential (and hence the force) between two atoms. No importance has been placed on the choice of values for the coefficients A and B of the attractive and repulsive forces, thus the *magnitudes* of the resulting forces have no physical significance. In order to be able to easily relate a force to an atomic relaxation (movement), the forces were normalised according to the force at the fixed distance of $2r_0$. Thus giving numbers of the order 1. Further control is provided by the relaxation constant 'α', which in these early examples was set equal to 1 Å. The latter implies that the atoms move distances of about 1 Å or so, in any one iteration of the positions.

3. As the repulsive force increases without limit, as the separation between two atoms decreases, two additional lines were added to the computer code immediately after the calculation of the response (dx and dy) of the atom in question. These lines put an upper limit on the potential movement of the atom, and prevent two atoms flying apart rapidly after their *initially* chosen random positions placed them close together.

8.5 INITIAL SIMULATIONS AND CONVERGENCE TESTS

Fig. 8.8 displays the results of simulations of 12 identical atoms, with circularly symmetric interaction potentials, starting from the initial positions illustrated in the previous figure.

Whilst it can be seen that the movement of the atoms in the first 100 iterations is substantial, by that point the atoms are already beginning to show the same symmetry as the 'final' positions obtained after 1000 iterations. All four simulations exhibit hexagonal symmetry in the 'equilibrium' atomic positions, with a distance between any two atoms which is very close to the chosen equilibrium separation of 2 Å. Such hexagonal configu-rations are common in the solid state systems that the simulation mirrors—collections of identical atoms such as occur in the metallic elements[7].

Fig. 8.9 illustrates the effect that the parameter α has on the simulations. It can be seen that in all three cases ($\alpha = 0.5$, 1 and 2 Å) the atoms converge after 1000 iterations towards similar equilibrium configurations, i.e. the hexagonal structure. However, closer inspection shows that the $\alpha = 2$ simulation (where the atoms are allowed to move distances up to about 2 Å per iteration) is more regular than the others. Thus it would appear, that in this example, this value of α is preferable. The point to note is that the final converged

---

[7]See any of the solid state physics books referred to so far.

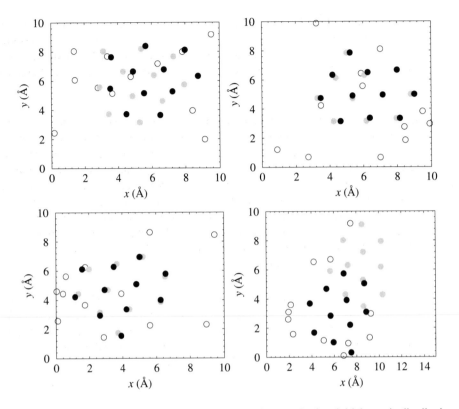

Fig. 8.8 The consequences of performing the simulations on the four initial atomic distributions of the previous figure. The initial positions are shown by the white circles, the positions after (I=) 100 iterations are shown in grey and the final positions after (I=) 1000 iterations are shown in black

equilibrium positions of the atoms should be independent of the *combination* of values chosen for the parameter α and the number of iterations.

8.6 LARGER ENSEMBLES AND REPRODUCING NATURE

Fig. 8.10 illustrates the results of two simulations of 64 atoms (with the same interaction potential as before). The long-range nature of the hexagonal ordering is now apparent. Note how the crystal will expand beyond the boundaries of the initial (randomly chosen) distribution if necessary. Perhaps the surprising (unforeseen) results illustrated by both simulations in Fig. 8.10 are the *deviations* from perfect symmetry. The deviations are likely to be more complex than merely missing atoms—see Task 8.2.

This emergence of vacancies within the crystal lattice is very interesting, because it is something that was never envisaged when this numerical method was created. Vacancies *do* occur in natural crystal lattices, so even this simple model is able to reproduce real

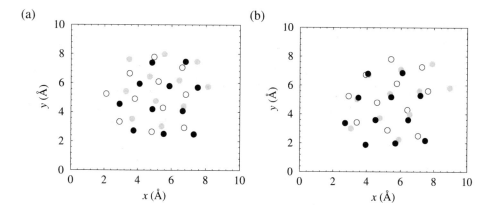

Fig. 8.9   The effect of the relaxation parameter α which connects the distance an atom moves to the total force upon it. The figures show the atomic positions after (a) I=100 iterations and (b) I=1000 iterations, for $\alpha = 0.5$ Å (white circles), $\alpha = 1$ Å (grey circles) and $\alpha = 2$ Å (black circles)

effects. However, the frequency at which vacancies arise in these molecular dynamic simulations is unlikely to be the same as that which occurs in nature, so any inference along those lines is likely to be dubious[8].

Such end results as displayed in Fig. 8.10 represent 'quasi-equilibrium' configurations, because they only represent a local minimum in the total potential energy—a lower state would surely be obtained by removing the vacancy. The calculation of the total energy of a molecular or crystal structure is often the aim of more sophisticated molecular dynamic simulations but, as mentioned already, in this introductory model the total potential energy can have no significance attached to it because of the arbitrary choice of the interaction potential energy (see equation (8.1)).

More sophisticated molecular dynamics simulations introduce the idea of a 'simulated anneal'. As the name suggests, it is a computational procedure developed to represent heating the atoms up slightly midway through the simulation, with the aim of 'escaping' from local energy minima and moving towards the true minimum. Such a procedure could be implemented within this scheme. After a certain number of iterations, the atoms could be jiggled a little by adding a small random displacement to their quasi-equilibrium positions. The simulation could then be allowed to proceed along the usual lines, with the atoms being allowed to interact and relax again into (possibly) new equilibrium positions. The simulation as developed here has been useful to illustrate several aspects of molecular dynamics and has served its purpose at this introductory level and, as such, the simulated anneal is not going to be developed during this text. The latter is, however, the subject of Project 8.3.

---

[8]Nature also produces the opposite of a vacancy—additional atoms between the usual lattice sites, so called 'interstitials'. This effect is unlikely to be reproduced in this work because of the simple spherical symmetry of the interatomic interaction potentials. Interstitials sit at local potential energy minima within the regular lattice and these will only occur if more complex angular dependent interaction potentials are employed.

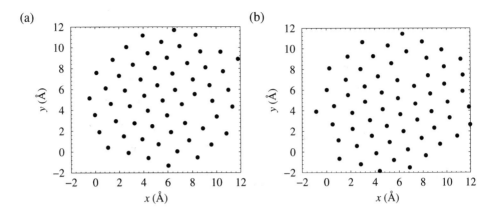

Fig. 8.10 Two 64 atom simulations ($\alpha = 1$, 1000 iterations and seed values of 1 and 2), showing the emergence of vacancies

8.7 NON-SPHERICAL INTERACTION POTENTIALS

Although the spherically symmetric interaction potentials have proved very useful in illustrating many points of molecular dynamics simulations, they really only represent simple metallic atoms. The wealth of other compounds which are so important in today's society are derived from a whole host of other atoms which require more complicated potentials for their description.

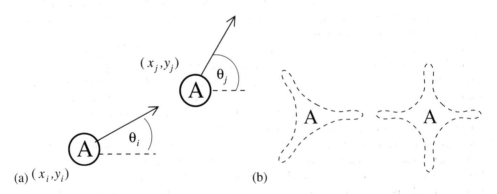

Fig. 8.11 (a) The angles involved with non-spherical (but still planar) potentials, and (b) examples of potentials with 3- and 4-fold symmetry

The latter almost certainly implies the abandonment of spherical symmetry and one way to achieve this (under the simplifying restriction of two dimensions) would be to replace the interaction potential $V(r)$:

$$V(r) \rightarrow \left| \cos\left(\frac{m\theta}{2}\right) \right| V(r) \qquad (8.10)$$

where m is the degree of the planar symmetry, and θ is the difference in the orientations of the atoms, i.e. $\theta = \theta_j - \theta_i$, see Fig. 8.11. Consideration of the cosine factor shows that, whenever $m\theta/2$ is a multiple of π, the interaction potential has a maximum and midway between these values it has a minimum, in this case zero. Fig. 8.11(b) illustrates this schematically for values of $m = 3$ and 4. The former could be used to represent the sp^2 hybridised bonds of carbon.

Task 8.4 discusses in more detail an initial simulation using a non-spherical interaction potential, but without the rotational degree of freedom of the atoms. Project 8.4 builds upon this but with the additional degree of freedom that the initial orientations of the atoms are random.

8.8 ORGANIC MOLECULES

Organic molecules[9] are the building blocks of life and the major components of the pharmaceutical and petrochemical industries and therefore constitute a class of compounds which are as important as semiconductors.

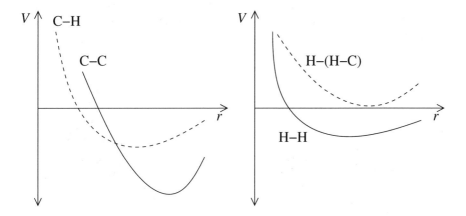

Fig. 8.12 Schematic illustrations of interaction potentials that may be required for simulations of hydrocarbons

Organic molecules have a vast range of sizes, from just a few to several hundreds of atoms and consequently they can have extremely complicated physical structures. Furthermore, the chemical (or therapeutic) properties may be dictated by just a few atoms which make up an 'active site'. The understanding of the structure (how and where the

---

[9]Molecules which involve carbon (C) and usually at least one hydrogen (H) atom, not CO_2, CO, CN.

atoms are joined together) can lead to the prediction of new compounds with possibly new desirable attributes. The latter has been a driving force behind the development of molecular dynamic simulations in this research field.

The wide variety of constituent elements, usually with complicated (non-spherical) interaction potentials, does make serious simulations of these materials beyond the level of an undergraduate textbook. However, some important features, relevant to the simulation of organic molecules, can be appreciated with some simple considerations.

For example, it is known that methane, which has the chemical symbol of CH_4, consists of a single carbon atom bonded to 4 hydrogen atoms. There are thus two interaction potentials that appear relevant:

1. The carbon–hydrogen (C–H) bond

2. The hydrogen–hydrogen (H–H) bond

It would be expected that both of these interactions would have potentials of the form, as previously shown, in Fig. 8.2, though of course the equilibrium separations would be different, so compared side by side they might look more like those illustrated in Fig. 8.12. The figure also gives a guessed interaction potential for the C–C bond—it is deeper, suggesting a stronger bond than H–H, and the equilibrium separation is greater because of the increased size of the atoms.

However, if a molecular dynamic simulation was performed as before, then it can be envisaged that although the 4 H atoms will bond to the C, they will also cluster around one side of the C in order to reduce their total energy through the attractive H–H interaction, see Fig. 8.13(a).

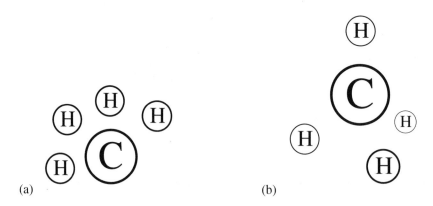

(a)   (b)

Fig. 8.13   (a) A possible result of a simple simulation of methane (CH_4), and (b) the actual structure of methane, a single carbon atom with 4 hydrogen atoms at the corners of a tetrahedron

Now, it is known that methane actually has the structure represented in Fig. 8.13(b), and the reason for the discrepancy is quite straightforward. When a hydrogen atom bonds to another type of atom, such as carbon (as in this case) or oxygen (as in water), the orbiting electron in the hydrogen atom is dragged over to one side to form the bond. As hydrogen has only one electron, then the positively charged nucleus is left bare, consequently a hydrogen atom in this bonded state will repel another hydrogen atom.

This has to be accounted for in a molecular dynamics simulation by choosing interaction potentials which account for the *environments* of the constituent atoms. For example, if both hydrogen atoms are already bonded to a carbon atom, then the hydrogen–hydrogen interaction will be repulsive, rather than attractive and the H–H potential in Fig. 8.12 will need to be replaced with something more like the H–(H–C) example. See the Tasks and Projects at the end of the chapter for examples on implementing calculations for some simple organic molecules.

Summary

The interaction between two atoms can be represented by a spherically symmetric potential of the form:

$$V = -\frac{A}{r^m} + \frac{B}{r^n}$$

where the first term represents the attractive and the second term the repulsive components. The indices m and n are often taken to have values '1' and perhaps '8' respectively. In the first instance the coefficients A and B can be chosen arbitrarily. However, their ratio can be fixed by recourse to the equilibrium separation r_0 of the atoms:

$$r_0^{-m+n} = \frac{nB}{mA}$$

If the separation of the jth atom from the ith atom is r_{ij} then the force due to this interaction potential V is given by:

$$\mathbf{F}_{ij} = -\left.\frac{\partial V}{\partial r}\right|_{r_{ij}}$$

It has been shown that one way to construct a molecular dynamic simulation of the interactions between a collection of atoms, is to calculate the total force on any one atom, due to its interactions with all the other atoms:

$$\mathbf{F}_i^{\text{total}} = \sum_{j=1, j \neq i}^{N} \mathbf{F}_{ij}(r_{ij})$$

and then allow that atom to move a small distance in the direction of the net force. This can be repeated many times for all the atoms, until an equilibrium is established. This final position can be taken as the lowest energy configuration and hence the atomic positions represent either a crystal or molecular structure.

Tasks

8.1 **Computational time:** Repeat the simulation of the metallic element introduced as the collection of identical atoms in Section 8.4, for increasing large numbers of atoms. Plot the computational time (for a fixed number of iterations) versus the number of atoms. Is there a relationship between the two? If there is, is it possible to deduce this by studying the structure of the algorithm and its implementation in computer code?

8.2 **Relaxation around a vacancy:** Section 8.6 illustrates simulations with 64 atoms, and shows occasions on which a vacancy (a missing atom) arises within the otherwise regular crystal lattice. Repeat these simulations, with more atoms if necessary, and with the computational parameters chosen in order to achieve as *high accuracy* as possible. The latter may be defined in terms of the regularity of the final array. Compare the regularity of the atomic positions around a vacancy with those in the rest of the crystal as a whole. *Note, more than 64 atoms may be necessary in order that the vacancy has a good probability of occurring towards the centre of the crystal and away from the influence of possible edge effects.*

8.3 **Impurity:** Substitute one atom from a collection of identical atoms 'A' with a different species 'B', say. Choose the interaction potential between an atom of type A and an atom of type B such that the equilibrium separation is twice that of the equilibrium separation between two atoms of type A. Make several simulations of this system and observe the nature of the crystal structure around the impurity atom.

8.4 **Interatomic potentials with 4-fold symmetry:** By choosing a suitable interatomic interaction potential, perform illustrative two-dimensional simulations of systems of identical atoms displaying 4-fold symmetry in their bonding. In this task, do not allow the atoms to rotate. For this additional degree of freedom, see Project 8.4.

8.5 **Two-dimensional (planar) methane calculation:** Choose the coefficients A and B for the attractive and repulsive components of spherically symmetric interaction potentials (of the form in equation (8.1) with indices $n = 8$ and $m = 1$, as before) in order to obtain appropriate potentials to represent the C–H and the H–H bonds, as in Fig. 8.12. Adjust the values of A and B to give a repulsive potential to represent the H–H interaction when either or both of the hydrogen atoms are bonded to carbon.

Randomly generate the positions of 1 carbon and 4 hydrogen atoms within a two-dimensional square box of side 10 Å, and perform a molecular dynamics simulation whilst retaining the restriction of only allowing the atoms to move within the two-dimensional plane. Give examples of equilibrium configurations of this 5-atom system and discuss.

8.6 **XYZ format and three-dimensional molecular viewers:** XYZ is a standard file format for recording atom types and their positions. Find its description by searching on the internet and adapt the simulation presented in this chapter to output the atomic positions and type in this format. There are three-dimensional molecular viewing software programs available in the public domain. Install one on your computer and demonstrate its capability, using some existing data.

Projects

8.1 **Three-dimensional simulations:** Adapt the two-dimensional simulation developed in Section 8.4 to three dimensions. Using suitable numbers of atoms, investigate the *stacking* properties of the layers.

8.2 **Nearest neighbours:** Simulations of very large numbers of atoms can take inordinate amounts of time to complete. One way to reduce this would be to consider only the interactions between the small number of neighbouring atoms. By starting from a hexagonal close-packed layer (the equilibrium points reached in Section 8.5), show the result of substituting one atom for a larger impurity, by considering the interaction of nearest neighbours only. Compare this with the results of Task 8.3.

8.3 **Removal of lattice defects with a simulated anneal:** Implement the simulated anneal, as discussed in Section 8.6, and apply it to the same 64 atom simulations. How large does the additional random movement of the atoms have to be in order to remove the vacancy?

8.4 **4-fold symmetry and rotational degrees of freedom**: Repeat the simulations described in Task 8.4. However, start with random orientations of the atoms as well as the usual random positions. Extend the molecular dynamics simulation by substituting the interatomic forces by torques. The latter may be implemented simply, by considering the response of an atom as a movement plus a rotation.

8.5 **Three-dimensional methane calculation:** Repeat Task 8.5, but this time generate the initial positions randomly within a three-dimensional box of side 10Å and allow the atoms to move in all directions. Make any necessary adjustments and alterations to the interaction potentials and/or molecular dynamics algorithm in order to try to represent methane realistically.

Appendix A:
FORTRAN implementation
of the shooting method

```
      PARAMETER(hbar=1.05459e-34)
      PARAMETER(e_0=1.602189e-19)

C      dE the energy increment
C dz the step length delta z
C E the energy E
C psi0,psi1,psi2 psi(z-dz), psi(z), and psi(z+dz)
C z the spatial coordinate
C
      REAL dE
      REAL dz
      REAL E
      REAL m
      REAL psi0,psi1,psi2
      REAL z
C
      dE=1e-3*e_0
      dz=1e-10
      m=9.109534e-31
C
```

```
      DO 2,E=0,e_0,dE
C        Set starting values
         psi0=0
         psi1=1
C        First value of z is delta z
         DO 1,z=dz,100e-10,dz
          psi2=(2*m*(dz/hbar)*(dz/hbar)*
     &          (e_0*(z/100e-10)*(z/100e-10)-E)+2)*psi1-psi0
          psi0=psi1
          psi1=psi2
    1    CONTINUE
         write(*,'(A,F11.6,A,E12.6)')'E=',E/(1e-3*e_0),
     &                               'meV psi(infty)=',psi2
    2 CONTINUE
      END
```

There are obviously some slight syntax changes, some of which are more obvious than others. One to be aware of is the definition of the mass m as a REAL rather than as a parameter. This is because FORTRAN assumes that variables beginning with the character m are integers.

Appendix B:
∇^2 *in spherical polar coordinates*

In spherical polar coordinates, the spatial dimension r is the distance of a point from the origin and can be found in terms of the Cartesian coordinates using Pythagoras's theorem:

$$\therefore r = \sqrt{x^2 + y^2 + z^2} \tag{B.1}$$

Now in spherically symmetric potentials, the wave function ψ is also spherically symmetric. Therefore, consider evaluating its derivates along the Cartesian axes:

$$\frac{\partial}{\partial x}\psi(r) = \frac{\partial}{\partial r}\psi(r) \times \frac{\partial r}{\partial x} \tag{B.2}$$

Differentiating both sides of equation (B.1) with respect to x, gives:

$$\frac{\partial r}{\partial x} = \frac{1}{2}\left(x^2 + y^2 + z^2\right)^{-\frac{1}{2}} \times 2x = \frac{x}{r} \tag{B.3}$$

Hence:

$$\frac{\partial}{\partial x}\psi(r) = \frac{\partial}{\partial r}\psi(r) \times \frac{x}{r} \tag{B.4}$$

The second derivative is then:

$$\frac{\partial}{\partial x}\frac{\partial}{\partial x}\psi(r) = \frac{\partial}{\partial x}\left[\frac{\partial}{\partial r}\psi(r) \times \frac{x}{r}\right] \tag{B.5}$$

$$\therefore \frac{\partial^2}{\partial x^2}\psi(r) = \frac{\partial^2}{\partial r^2}\psi(r) \times \frac{\partial r}{\partial x}\frac{x}{r} + \frac{\partial}{\partial r}\psi(r)\frac{\partial}{\partial x}\left(\frac{x}{r}\right) \tag{B.6}$$

and thus:

$$\frac{\partial^2}{\partial x^2}\psi(r) = \frac{x^2}{r^2}\frac{\partial^2}{\partial r^2}\psi(r) + \frac{\partial}{\partial r}\psi(r)\left(\frac{1}{r} - \frac{x}{r^2}\frac{\partial r}{\partial x}\right) \tag{B.7}$$

Finally:

$$\frac{\partial^2}{\partial x^2}\psi(r) = \frac{1}{r}\frac{\partial}{\partial r}\psi(r) - \frac{x^2}{r^3}\frac{\partial}{\partial r}\psi(r) + \frac{x^2}{r^2}\frac{\partial^2}{\partial r^2}\psi(r) \tag{B.8}$$

Similar expressions follow for the equivalent y- and z-axes. Hence, the complete expression for $\nabla^2\psi(r)$ is given by:

$$\left(\frac{\partial^2}{\partial x^2} + \frac{\partial^2}{\partial y^2} + \frac{\partial^2}{\partial z^2}\right)\psi(r) =$$

$$\frac{3}{r}\frac{\partial}{\partial r}\psi(r) - \frac{(x^2+y^2+z^2)}{r^3}\frac{\partial}{\partial r}\psi(r) + \frac{(x^2+y^2+z^2)}{r^2}\frac{\partial^2}{\partial r^2}\psi(r) \tag{B.9}$$

which gives:

$$\left(\frac{\partial^2}{\partial x^2} + \frac{\partial^2}{\partial y^2} + \frac{\partial^2}{\partial z^2}\right)\psi(r) = \frac{2}{r}\frac{\partial}{\partial r}\psi(r) + \frac{\partial^2}{\partial r^2}\psi(r) \tag{B.10}$$

(as, for example, in Weidner and Sells, *Elementary Modern Physics*, (Allyn and Bacon, Boston, 1980) Third Edition, p. 188).

Appendix C:
A comment on the computer
sourcecodes

The computer codes developed as part of the book were created with the ideas of simplicity and ease of understanding as paramount. They could be slicker and quicker, but this would be at the expense of transparency.

Some of the codes were unavoidably long, so to spare the reader the tedious task of retyping them, they can be downloaded from the book's web site, see:

```
http://www.imp.leeds.ac.uk/CM/
```

Appendix D:
Note for tutors

Hints and skeletal answers to the problems are available from the book's website:

`http://www.imp.leeds.ac.uk/CM/`

If further assistance is required with any particular problem, then please don't hesitate to contact the author.

References

1. J. W. Leech, *Classical Mechanics*, Chapman and Hall, London, Second edition, 1965.

2. R. T. Weidner and R. L. Sells, *Elementary Modern Physics*, Allyn and Bacon, Boston, Third edition, 1980.

3. R. M. Eisberg, *Fundamentals of Modern Physics*, Wiley, New York, 1961.

4. P. A. M. Dirac, *The Principles of Quantum Mechanics*, Clarendon Press, Oxford, Fourth edition, 1967.

5. Milton Abramowitz and Irene A. Stegun, *Handbook of Mathematical Functions*, Dover Publications Inc., New York, 1965.

6. M. J. Kelly, *Low Dimensional Semiconductors: Materials, Physics, Technology, Devices*, Clarendon Press, Oxford, 1995.

7. J. P. Killingbeck, *Microcomputer Algorithms*, Hilger, Bristol, 1992.

8. L. I. Schiff, *Quantum Mechanics*, McGraw-Hill, London, 1968.

9. D. K. Ferry, *Quantum Mechanics: An Introduction for Device Physicists and Electrical Engineers*, IOP Publishing, London, 1995.

10. P. Harrison, *Quantum Wells, Wires and Dots: Theoretical and computational physics*, John Wiley and Sons, Ltd, Chichester, United Kingdom, 1999, 456 pages, ISBN 0 471 98495 7.

11. R. L. Liboff, *Introductory Quantum Mechanics*, Holden-Day, San Francisco, 1980.

12. J. P. Killingbeck, *Microcomputer Quantum Mechanics*, Adam Hilger, Bristol, 1983.

13. E. Anderson, Z. Bai, C. Bischof, J. Demmel, J. Dongarra, J. Du Croz, A. Greenbaum, S. Hammarling, A. McKenney, S. Ostrouchov, and D. Sorensen, *LAPACK Users' Guide*, Society for Industrial and Applied Mathematics, Philadelphia, Second edition, 1995.

14. J. Crank, *The Mathematics of Diffusion*, Oxford University Press, London, 1956.

15. W. D. Callister Jr., *Materials Science and Engineering*, John Wiley and Sons, Inc., New York, 1985.

16. D. Stauffer and A. Aharony, *Introduction to Percolation Theory*, Taylor and Francis, London, Second edition, 1994.

17. J. S. Blakemore, *Solid State Physics*, University Press, Cambridge, Second edition, 1985.

18. S. O. Kasap, *Principles of Electrical Engineering Materials and Devices*, Irwin McGraw-Hill, Boston, 1997.

19. N. W. Ashcroft and N. D. Mermin, *Solid State Physics*, Saunders College Publishing, Philadelphia, 1976.

Index